PHILIP LUTLEY SCLATER

ARGENTINE ORNITHOLOGY

A
DESCRIPTIVE CATALOGUE
OF THE
BIRDS OF THE ARGENTINE REPUBLIC

WITH NOTES ON THEIR HABITS
BY
W.H. HUDSON

VOLUME I

Elibron Classics
www.elibron.com

Elibron Classics series.

ISBN 1-4021-0174-0 (paperback)
ISBN 1-4021-0173-2 (hardcover)

This Elibron Classics Replica Edition is an unabridged facsimile of the edition published in 1888 by R.H. Porter, London.

ARGENTINE ORNITHOLOGY.

A
DESCRIPTIVE CATALOGUE
OF THE
BIRDS OF THE ARGENTINE REPUBLIC.

BY
P. L. SCLATER, M.A., PH.D., F.R.S., ETC.

WITH NOTES ON THEIR HABITS

BY
W. H. HUDSON, C.M.Z.S.,
LATE OF BUENOS AYRES.

THE CARIAMA.

VOLUME I.

LONDON:
R. H. PORTER, 6 TENTERDEN STREET, W.
1888.

ALERE FLAMMAM.

PRINTED BY TAYLOR AND FRANCIS,
RED LION COURT, FLEET STREET.

PREFACE TO THE FIRST VOLUME.

THE present volume contains an account of the Passeres of the Argentine Republic, which, as at present known, number some 229 species. The second volume, which it is hoped will be ready in the course of next year, will be devoted to the history of the remaining Orders of Birds, and will also contain the Introduction and Index, and complete the work.

All the personal observations recorded in these pages are due to Mr. Hudson, while I am responsible for the arrangement, nomenclature, and scientific portions of the work.

I have to acknowledge with many thanks a donation of £40 from the Royal Society, which has enabled Mr. Hudson to devote a portion of his time to the compilation of his interesting notes.

P. L. S.

December 1, 1887.

CONTENTS OF VOL. I.

Fam. X. FRINGILLIDÆ, or Finches.

Fam. XVIII. FORMICARIIDÆ, or ANT-BIRDS.

Fam. XIX. PTEROPTOCHIDÆ, or TAPACOLAS.

LIST OF PLATES IN VOL. I.

ARGENTINE ORNITHOLOGY.

Order I. PASSERES.

Suborder I. *OSCINES.*

Fam. I. TURDIDÆ, or THRUSHES.

Eight species of the almost cosmopolitan Thrush-family are known to occur in the Argentine Republic. Of these, five belong to the widely-spread genus *Turdus,* and are closely similar in structure to our European Thrushes. The remaining three are Mock-birds (*Mimina*)—a group restricted to the New World, and sometimes considered more nearly allied to the Wrens. They are remarkable as splendid songsters. Burmeister includes in his list a fourth species of *Mimus* (*M. thenca*), as found in the Mendoza district; but there may be some error in this, as *M. thenca* is only known to us from Chili west of the Andes.

1. TURDUS LEUCOMELAS, Vieill.

(DUSKY THRUSH.)

Turdus leucomelas, *Scl. et Salv. Nomencl.* p. 1; *Hudson, P. Z. S.* 1870, p. 798 (Buenos Ayres); *Durnford, Ibis,* 1877, p. 166 (Buenos Ayres); *White, P. Z. S.* 1882, p. 592 (Misiones and Corrientes); *Barrows, Bull. Nutt. Orn. Cl.* viii. p. 85 (Concepcion); *Seebohm, Cat. Birds,* v. p. 213. **Turdus crotopezus,** *Burm. La-Plata Reise,* ii. p. 474 (Mendoza).

Description.—Above olive-grey, tinged with brown on the head and neck; beneath pale grey, throat white, more or less striped with brown; middle of belly and crissum white; under wing-coverts and inner margins of wing-feathers fulvous; bill yellow; feet hazel: total length 9·0 inches, wing 4·5, tail 3·7. *Female* similar.

Hab. Eastern South America, from Cayenne to Buenos Ayres.

The Dusky Thrush is widely distributed in South America, and ranges as far south as Buenos Ayres, where it is quite common in the woods along the Plata river. It is a shy forest-bird; a fruit- and insect-eater; abrupt in its motions; runs rapidly on the ground with beak elevated, and at intervals pauses and shakes its tail; pugnacious in temper; strong on the wing, its flight not being over the trees, but masked by their shadows. It can always be easily distinguished, even at a distance, from other species by its peculiar short, metallic chirp—a melodious sound indicating alarm or curiosity, and uttered before flight—in contrast to the harsh screams and chuckling notes of other Thrushes in this district.

Whether it is a fine singer or not within the tropics I am unable to say, its vocal powers having received no attention from the naturalists who have observed it. With us in the temperate climate of Buenos Ayres, where it commences to sing in September, it has the finest song of any bird I know, excepting only *Mimus triurus*. Like the English Song-Thrush, but unlike its near neighbours the Red-bellied Thrush and the Magellanic Thrush, it perches on the summit of a tree to sing. Its song is, however, utterly unlike that of the English bird, which is so fragmentary, and, as Mr. Barrows describes it, made up of "vocal attitudes and poses." The two birds differ also in voice as much as in manner. The strains of the Dusky Thrush are poured forth in a continuous stream, with all the hurry and freedom of the Sky-Lark's song; but though so rapidly uttered, every note is distinct and clear, and the voice singularly sweet and far-reaching. At intervals in the song there recurs a two-syllabled note twice repeated, unlike in sound any other bird-music I have heard, for it is purely metallic, and its joyous bell-like "*te-ling te-ling*" always comes like a delightful surprise to the listener, being in strange contrast with the prevailing tone.

The song is altogether a very fine one, its peculiar charm being that it seems to combine two opposite qualities of bird-music, plaintiveness and joyousness, in some indefinable manner.

I have never heard this species sing in a cage or anywhere near a human habitation; and it is probably owing to its recluse habits that its excellent song has not been hitherto noticed. Azara perhaps mistook the song of this species for that of *Turdus rufiventris*, a very inferior vocalist.

The nest is made in the centre of a thick bush or tree six or eight feet above the ground, and is a deep elaborate structure, plastered inside with mud, and lined with soft dry grass. The eggs are four in number, oblong; the ground-colour light blue, abundantly marked with reddish-brown spots.

This Thrush has, I believe, a partial migration in Buenos Ayres. In the autumn and winter I have frequently observed it in localities where it is never seen in summer.

---

2. TURDUS RUFIVENTRIS, Vieill.

(RED-BELLIED THRUSH.)

Turdus rufiventris, *Scl. et Salv. Nomencl.* p. 2; *Burm. La-Plata Reise*, ii. p. 474; *Hudson, P. Z. S.* 1870, p. 89 (Buenos Ayres); *Durnford, Ibis*, 1877, p. 167, 1880, p. 417 (Buenos Ayres); *White, P. Z. S.* 1882, p. 592 (Salta); *Barrows, Bull. Nutt. Orn. Cl.* viii. p. 85 (Concepcion). **Turdus rufiventer,** *Seebohm, Cat. B.* v. p. 222.

Description.—Entire upper parts olive-grey; throat white, striped with dark brown, the striped portion extending to the chest; rest of under surface, also under wing-coverts, rufous-red, deepest on the belly; bill brownish yellow; feet brown: total length 9·0 inches, wing 4·6, tail 3·6. *Female* similar.

Hab. S.E. Brazil, Paraguay, Uruguay, and La Plata.

The Red-bellied Thrush, distinguished from the species just described by its larger size and the bright rufous colour of its under plumage, is common everywhere in the Plata region, and does not appear to be migratory. It is a noisy, strong-winged, quarrelsome bird, closely resembling the Dusky Thrush in its manners. It inhabits forests, runs on the ground in search of food, and when approached darts away with loud chuckling notes, flying close to the ground. These birds are also often seen pursuing each other through the trees with loud harsh screams.

The song has a faint resemblance to that of the English Song-Thrush, being composed of a variety of notes uttered in the same disconnected manner, with frequent pauses; but it is, both in sweetness and strength, inferior to that of the English bird. As a rule this Thrush sings concealed in a thick bush or tree.

The nest is deep, well made, plastered inside with mud, and concealed in the centre of a large bush or low tree. The eggs are four, pale blue in colour, and thickly spotted with brown.

---

3. TURDUS MAGELLANICUS, King.

(MAGELLANIC THRUSH.)

Turdus magellanicus, *Seebohm, Cat. B.* v. p. 223, pl. xiv. **Turdus falklandicus,** *Scl. et Salv. Nomencl.* p. 2; *Hudson, P. Z. S.* 1872, p. 548 (Rio Negro); *Döring, Exp. al Rio Negro, Zool.* p. 36 (R. Colorado and R. Negro).

Description.—Head, wings, and tail brownish black; the remaining portions of upper surface brownish olive; under surface pale rufous, obscure on the breast; throat white, striped with black; bill and feet dull yellow: total length 10·5 inches, wing 7·5, tail 4·0. *Female* similar.

Hab. Patagonia and Chili.

This Thrush, which I met with on the Rio Negro of Patagonia, so closely approximates to *T. rufiventris* in colour, language, and habits, that the casual observer finds it difficult to distinguish one from the other. Its nest and eggs are also precisely like those of its northern representative. The song is, however, much poorer, and reminds one of the first attempts of a young bird. That a member of so melodious a family should have so inferior a song I attribute to the fact that Thrushes (unlike the songsters of other genera) sing only in the warm season and when the air is calm. In the southern portion of the South-American continent violent winds prevail in summer, so that this southern Thrush sings perhaps less frequently than any other song-bird, and appears to be losing the faculty of song altogether.

---

4. TURDUS FUSCATER, d'Orb. et Lafr.

(ARGENTINE BLACKBIRD.)

Turdus fuscater, *Scl. et Salv. Nomencl.* p. 2; *White, P. Z. S.* 1882, p. 593 (Catamarca); *Burm. La-Plata Reise*, ii. p. 474 (Mendoza, Cordova, and Tucuman).
Merula fuscatra, *Seebohm, Cat. B.* v. p. 243.

Description.—Uniform brownish black; bill and feet yellow: total length 11·5 inches, wing 5·3, tail 4·5. *Female* similar.

Hab. Bolivia and Northern La Plata.

"I have noticed this bird in the provinces of Mendoza, San Juan, San Luis, Cordova, and Catamarca, where it is much esteemed as a cage-bird on account of its beautiful whistle. In Mendoza it goes by the name of 'Zorsal,' in Catamarca and Cordova by that of 'Merlo,' and in Andalgala (Catamarca province) 'Chocoyno.'"—*White, l. s. c.*

"Very common at Mendoza, Cordova, and Tucuman, and met with among the shady trees of the promenades: has a better song than our Blackbird, which is heard at a distance, especially in the evening. Called 'Crispin' in Tucuman."—*Burm. l. s. c.*

---

5. TURDUS NIGRICEPS, Cab.

(BLACK-HEADED THRUSH.)

Turdus nigriceps, *Cab. J. f. O.* 1878, p. 195.

Description.—Above slaty grey, head black; below paler, chin white, striped with black; middle of belly and crissum white; bill and feet pale yellow: total length 7·6 inches, wing 4·4, tail 3·5. *Female* olive-brown, paler below, chin pale ochraceous; middle of belly and crissum white; bill brown, feet yellowish.

Hab. Peru, Bolivia, and Northern La Plata.

Met with by Dr. A. Döring and Herr Schulz on the Sierra of Cordova.

---

6. MIMUS MODULATOR, Gould.

(CALANDRIA MOCKING-BIRD.)

Mimus modulator, *Sharpe, Cat. B.* vi. p. 347. **Mimus calandria,** *Burm. La-Plata Reise,* ii. p. 475; *Scl. et Salv. Nomencl.* p. 3; *Hudson, P. Z. S.* 1870, p. 89 (Buenos Ayres); *Durnford, Ibis,* 1877, p. 167 (Buenos Ayres); *White, P. Z. S.* 1883, p. 37 (Cordova); *Barrows, Bull. Nutt. Orn. Cl.* viii. p. 85 (Concepcion).

Description.—Above dark grey, faintly mottled with pale grey; rump tinged with brown; wings nearly black, the feathers edged with brownish white; tail black, the feathers, except the two middle ones, broadly tipped with white; under surface dirty white; bill and legs black; eye olive-green: total length 11·0 inches, wing 4·5, tail 4·9. *Female* similar.

Hab. Paraguay, Uruguay, and Argentine Republic.

Azara has not failed to remark that it would be well to find a more appropriate name for this species, which was absurdly called "Calandria" (*i. e.* Sky-Lark) by the early colonists of the Plata. Use is, however, too strong to be easily set aside, and the name is now familiar to everyone in the Argentine Province. Moreover, by a curious irony of fate, the Spanish naturalist himself, by employing this unsuitable name in his 'Apuntamientos,' even while protesting against it, has been the cause of its introduction into scientific nomenclature.

It would be impossible to improve on the account Azara gives of the bird's appearance and manners. The prevailing colour of the plumage is grey, the irides are deep green, the beak black, slender, and curved. The tail is long, jerked and elevated when the bird is at rest, spread open and depressed in flight. The Calandria's movements are measured and dignified, its flight low and never extends far, the bird usually passing from one tree to another in a long graceful curve. It goes alone or with its mate only; feeds chiefly on the ground; does not penetrate into deep forests, nor is it seen on the treeless plains. It frequents the borders of woods and open grounds abounding in isolated shrubs and trees; is fond of coming about houses, and invariably perches itself on the most conspicuous places. It sings chiefly in spring, and its really wonderful vocal powers have made it one of our best-known and most-admired songsters. To sing it usually places itself on the summit of a bush or tree, and occasionally, as if carried away by excitement, it darts upwards three or four yards into the air, and then drops back on to its perch. So varied are its notes, and so frequently suggestive of the language of other species, that the listener finds himself continually asking whether the Calandria is really an original singer or merely a cunning plagiarist, able to steal scraps of fifty different melodies and to blend them in some sort into one complete composition. As a whole the song is in character utterly unlike that of any other bird (birds of the *Mimus* genus, of course, excepted), for the same notes are never repeated twice in the same order; and though the Calandria has many favourite notes, he is able to vary every one of them a hundred ways. Sometimes the whole song seems to be made up of imitations of other singers, with slight variations—and not of singers only, for now there will be clear flute-like notes, only to be succeeded by others reedy and querulous as the hunger-calls of a young Finch; then there will be pretty flourishes or Thrush-like phrases, and afterwards screams, as of a frightened Swallow hurrying through the sky to announce the approach of a Falcon; or perhaps piteous outcries, as of a chicken in the clutches of a Kite.

Nevertheless Azara says truly that the Calandria does not mock or mimic the songs of other birds; for though the style and intonation of a score of different singers, chatterers, and screamers are reproduced by him, one can never catch a song, or even a portion of a song, of which he is able to say that it is absolutely like that of any other species. This much, however, can be said of the Calandria: he has a passion for endless variety in singing, a capacity for varying his tones to almost any extent, and a facility for catching the notes of other birds, which, in the Virginian Mocking-bird of North, and in the White-banded Mocking-bird of South America, has been developed into that marvellous faculty these two species possess of faithfully imitating the songs of all other birds. The two species I have just named, while mockers of the songs of other birds, also retain their own original music—their "natural song," as an American ornithologist calls it.

The Calandria makes its nest in the middle of a large bush or low thorn-tree standing by itself; it is deep, like the nest of a Thrush in form, built of sticks, thorns, and grass, and lined with thistle-down or some other soft material. The eggs are four or five, pale blue, and thickly marked with reddish-brown spots.

When the nest is approached the parent birds demonstrate their anxiety by uttering loud harsh angry notes.

It is generally believed that the Calandria will not live in captivity. I have, however, seen a few individuals in cages, but they never sang.

7. MIMUS PATACHONICUS (d'Orb. et Lafr.).

(PATAGONIAN MOCKING-BIRD.)

Mimus patachonicus, *Scl. et Salv. Nomencl.* p. 3; *Hudson, P. Z. S.* 1872, p. 538 (Rio Negro); *Durnford, Ibis,* 1877, p. 31 (Chupat); *Döring, Exp. al Rio Negro, Zool.* p. 30 (R. Colorado); *Sharpe, Cat. B.* vi. p. 352. **Mimus thenca,** *Burm. La-Plata Reise,* ii. p. 475 (Mendoza)?

Description.—Above and beneath grey, paler on the under surface, and tinged with rufous on the belly; throat and mark over the eye white; wings black, the outer webs of the feathers edged with white and grey; tail black, tipped with white; bill and feet black; eye olive-green: total length 9·2 inches, wing 4·2, tail 4·0. *Female* similar, but smaller in size and lighter in colour.

Hab. Northern Patagonia.

The Patagonian Mocking-bird, which I met with during my sojourn on the Rio Negro of Patagonia in 1871, closely resembles the species just described, but is smaller, the plumage is of a darker grey, and the irides are also of a darker green. It is a common bird, resident, lives alone or with its mate, feeds on insects and berries, and in its manner of flight and habits is like *Mimus modulator*. The nest is made in the centre of a bush of thorns and sticks, and lined with dry grass, cow-hair, or other soft material; and the eggs are four in number, bluntly pointed, and thickly marked with dark flesh-coloured spots. When the nest is approached the parent birds come close to the intruder, often perching within a yard of his head, but without uttering any sound, differing in this respect from *M. modulator*.

The song of the Patagonian bird is in character like that of the northern species, the variety of its notes being apparently infinite; there are, however, some differences worth mentioning. The singing of the Patagonian species is perhaps inferior, his voice being less powerful, while his mellow and clear notes are constantly mingled with shrill ones, resembling the cries of some of the Dendrocolaptine birds. While incapable of notes so loud or so harsh as those of the northern bird, or of changes so wild and sudden, he possesses even a greater variety of soft notes. Day after day for many months I have heard them singing, yet never once listened to them for any length of time without hearing some note or phrase I had never heard before. The remarks I have made concerning the Calandria's mocking-faculties also apply to this bird: but though he does not actually repeat the notes and songs of other species, he certainly does mock the notes of individuals of his own species; for it must be borne in mind that no two individuals sing quite alike, and that the same bird constantly introduces new notes into his song, and never repeats his notes in the same order. I have often observed that when a bird, while singing, emits a few of these *new* notes, he seems surprised and delighted with them; for, after a silent pause, he repeats them again and again a vast number of times, as if to impress them on his memory. When he once more resumes his varied singing, for hours, and sometimes for days, the expression he has discovered is still a favourite one, and recurs with the greatest frequency. But this is not all. If the new note or phrase happens to be a very striking one, it immediately takes the fancy of all the other birds within hearing, and often in a small thicket there will be a dozen or twenty birds near together, each sitting perched on the summit of his own bush. After the new wonderful note has been sounded they all become silent and attentive, reminding one in their manner of a caged Parrot listening to a sound it is trying to learn. Presently they learn it, and are as pleased with its acquisition as if they had discovered it themselves, repeating it incessantly. I noticed this curious habit of the bird many times, and on one occasion I found that for three entire days all the birds in a small thicket I used to visit every day did nothing but repeat incessantly two or three singular notes which they had borrowed from one of their number. The constant repetition of this one sound had a strongly irritating effect on me; but a day or two later they had apparently got tired of it themselves, and had resumed their usual varied singing.

This bird usually sits still upon the summit of a bush when singing, and its music is heard in all seasons and in all weathers from dawn till after dark: as a rule it sings in a leisurely unexcited manner, remaining silent for some time after every five or six or a dozen notes, and apparently listening to his brother-performers. These snatches of melody often seem like a prelude or promise of something better coming; there is often in them such exquisite sweetness and so much variety that the hearer is ever wishing for a fuller measure, and still the bird opens his bill to delight and disappoint him, as if not yet ready to display his whole power.

---

8. MIMUS TRIURUS (Vieill.).

(WHITE-BANDED MOCKING-BIRD.)

[Plate I.]

Mimus triurus, *Scl. et Salv. Nomencl.* p. 3; *Hudson, P. Z. S.* 1872, p. 539 (Rio Negro); *White, P. Z. S.* 1882, p. 593 (Buenos Ayres); *Burm. La-Plata Reise,* ii. p. 475 (Mendoza, Cordova, and Tucuman); *Sharpe, Cat. B.* vi. p. 342.

Description.—Above grey, brown on the rump; beneath light grey, white on the belly; wing black, crossed with a broad white band; tail white, except the

MIMUS TRIURUS.

two middle feathers, which are black; bill and feet black; eye orange-yellow: total length 9·5 inches, wing 4·8, tail 4·2. *Female* similar.

Hab. Paraguay, Argentine Republic, and Bolivia.

Azara first met with this king of the Mocking-birds in Paraguay a century ago; he named it "Calandria de las tres colas," and described the plumage accurately, but was, I think, mistaken about the colour of the eye, which is orange-red and not olive-green. He says that it is a rare species, possessing no melodious notes, which proves at once that he never heard it sing. D'Orbigny obtained it in Bolivia, Bridges in Mendoza, and more recently it has been found by collectors in various parts of the Argentine country, even in Buenos Ayres, where, however, it is probably only an occasional visitor. But they have told us nothing of its song and of its miraculous mocking-powers. For my part I can think of no other way to describe the surpassing excellence of its melody, which delights the soul beyond all other bird-music, than by saying that this bird is among song-birds like the diamond among stones, which in its many-coloured splendour represents and exceeds the special beauty of every other gem.

I met with this species on the Rio Negro in Patagonia: it was there called *Calandria blanca*, a name not strictly accurate, since the bird is not all white, but certainly better than Azara's strange invention of "Lark with three tails."

The bird was not common in Patagonia, and its only language was a very loud harsh startled note, resembling that of the *Mimus calandria*; but it was past the love-season when I first met with it, and the natives all assured me that it possessed a very wonderful song, surpassing the songs of all other birds; also that it had the faculty of imitating other species. In manners and appearance it struck me as being utterly unlike a *Mimus*; in its flight and in the conspicuous white and black of the wings and tail, it looked like a Tyrant of the *Tænioptera* group. It was extremely shy, had a swift, easy, powerful flight, and, when approached, would rise up high in the air and soar away to a great distance. In February it disappeared from the Rio Negro and did not return till the following October, after the arrival of all the other migrants. It was then that I had the rare good fortune to hear it sing, and I shall never forget the sensation I experienced when listening to its matchless melody.

While walking through a *chañar*-wood one bright morning, my attention was suddenly arrested by notes issuing from a thicket close by, and to which I listened in delighted astonishment, so vastly superior in melody, strength, and variety did they seem to all other bird-music. That it was the song of a *Mimus* did not occur to me; for while the music came in a continuous stream, until I marvelled that the throat of any bird could sustain so powerful and varied a song for so long a time, it was never once degraded by the harsh cries, fantastical flights, and squealing buffooneries so frequently introduced by the Calandria, but every note was in harmony and uttered with a rapidity and joyous abandon no other bird is capable of, except, perhaps, the Sky-Lark; while the purity of the sounds gave to the whole performance something of the ethereal rapturous character of the Lark's song when it comes to the listener from a great height in the air.

Presently this flow of exquisite unfamiliar music ceased, while I still remained standing amongst the trees, not daring to move for fear of scaring away the strange vocalist. After a short interval of silence I had a fresh surprise. From the very spot whence that torrent of melody had issued, burst out the shrill, confused, impetuous song of the small yellow-and-grey Patagonian Flycatcher (*Stigmatura flavo-cinerea*). It irritated me to hear this familiar and trivial song after the other, and I began to fear that my entertainer had flown away unobserved. But in another moment, from the same spot, came the mellow matin-song of the Diuca Finch, and this was quickly succeeded by the silvery bell-like trilling song of the Churrinche, or little Scarlet Tyrant-bird. Then followed many other familiar notes and songs—the flute-like evening-call of the Crested Tinamou, the gay hurried twittering of the Black-headed Goldfinch, and the leisurely-uttered, delicious strains of the Yellow Cardinal, all repeated with miraculous fidelity. How much was my wonder and admiration increased by the discovery that my one sweet singer had produced all these diverse strains! The discovery was only made when he began to repeat songs of species that never visit Patagonia. I knew then that I was at last listening to the famed White Mocking-bird, just returned from his winter travels, and repeating in this southern region the notes he had acquired in subtropical forests a thousand miles away.

These imitations at length ceased, after which the sweet vocalist resumed his own matchless song once more. I ventured then to creep a little nearer, and at length caught sight of him not fifteen yards away. I then found that the pleasure of listening to its melody was greatly enhanced when I could at the same time see the bird, so carried away with rapture does he seem while singing, so many and so beautiful are the gestures and motions with which his notes are accompanied. He passes incessantly from bush to bush, scarcely alighting on their summits, and at times dropping down beneath the foliage; then, at intervals, soaring to a height of a hundred feet above the thicket, with a flight slow as that of a Heron, or mounting suddenly upwards with a wild,

CINCLUS SCHULZI

hurried, zigzag motion; then slowly circling downwards, to sit with tail outspread and the broad glistening white wings expanded, or languidly waved up and down like the wings of some great butterfly—an object beautiful to see.

When I first heard this bird sing I felt convinced that no other feathered songster on the globe could compare with it; for besides the faculty of reproducing the songs of other species, which it possesses in common with the Virginian Mocking-bird, it has a song of its own, which I believed matchless; and in this belief I was confirmed when, shortly after hearing it, I visited England, and found of how much less account than this Patagonian bird, which no poet has ever praised, were the sweetest of the famed melodists of the Old World.

Fam. II. CINCLIDÆ, or DIPPERS.

The Dippers, constituting the genus *Cinclus* and the family Cinclidæ, are sparingly distributed, principally in the Alpine Regions which contain clear and perennial streams, throughout the Palæarctic and Nearctic Regions. In the Neotropical Region they are represented by three species, one of which is found in the northern provinces of the Argentine Republic.

---

9. CINCLUS SCHULZI, Cab.

(SCHULZ'S DIPPER.)

[Plate II.]

Cinclus schulzi, *Cab. J. f. O.* 1883, p. 102, t. ii. fig. 3.

Description.—Dark grey; throat pale rufous; a broad bar on the inner webs of the wing-feathers white: total length 5·5 inches, wing 3·0, tail 1·6.

Hab. Northern Argentina.

A recent discovery of Herr Fritsch Schulz, who obtained specimens of it on the Cerro Vayo of Tucuman, where this species, like others of the genus, frequents the mountain-streams.

Fam. III. MUSCICAPIDÆ, OR FLYCATCHERS.

The peculiar genus *Polioptila*, which contains some twelve or thirteen species of small-sized American birds, ranging from the United States to the Argentine Republic, has been variously arranged by naturalists, but seems to be more nearly related to the African genus *Stenostira* than to any other known form. I therefore now place it with the Muscicapidæ, or Flycatchers, of which it is the only genus found in the New World.

---

10. POLIOPTILA DUMICOLA (Vieill.).

(BRUSH-LOVING FLY-SNAPPER.)

Polioptila dumicola, *Scl. et Salv. Nomencl.* p. 4; *Durnford, Ibis,* 1876, p. 157, 1877, p. 167 (Buenos Ayres); *Salv. Ibis,* 1880, p. 352 (Tucuman); *White, P. Z. S.* 1882, p. 593 (Buenos Ayres); *Barrows, Bull. Nutt. Orn. Cl.* viii. p. 86 (Concepcion); *Sharpe, Cat. B.* x. p. 444. **Culicivora dumicola,** *Burm. La-Plata Reise,* ii. p. 473 (Paraná). **Culicivora boliviana,** *Scl. P. Z. S.* 1852, p. 34, pl. xlvii.

Description.—Above clear greyish blue; wing-coverts, bastard-wing, and primary-coverts dusky brown, with greyish-blue edges; quills dusky; upper tail-coverts and tail-feathers black, the third outer tail-feather white at the tip, the outer tail-feather nearly entirely white; from the base of the forehead a black line extends backwards over the eye; under surface delicate ashy grey, white on the abdomen and under tail-coverts; bill and feet bluish black: total length 4·5 inches, wing 2·1, tail 2·0. *Female* similar, but without the black eye-streak.

Hab. Paraguay and Northern Argentina.

This little bird strongly resembles some species of that division of the Tyrannidæ which includes the genera *Stigmatura*, *Serpophaga*, and *Anæretes*; but the likeness, strange to say, is even more marked in habits and voice than in coloration and general appearance.

It is found in open thorny woods and thickets; and in Buenos Ayres seems to have a partial migration, as it is much more common in summer than in winter. At all times male and female are found together, and probably pair for life, like several of the species in the groups just mentioned. They are seen continually hopping about among the twigs in a leisurely deliberate manner, all the time emitting a variety of low short notes, as if conversing together; and at intervals they unite their voices in a burst of congratulatory notes, like those uttered by the small Tyrant-birds they resemble. They have no song. I have not found the nest, but Dr. Burmeister says that it is made in bushes, and that the eggs are white.

Fam. IV. TROGLODYTIDÆ, OR WRENS.

The Troglodytidæ, or Wren family, are of wide distribution, and are found alike under the tropics and in temperate latitudes. In South America nearly 100 species altogether are known to occur. Of these two are familiar inhabitants of the whole Argentine Republic, and a third, belonging to the water-loving genus *Donacobius*, is met with in the eastern provinces on the Paraná. A fourth species has been lately described from Tucuman.

---

11. DONACOBIUS ATRICAPILLUS (Linn.).
(BLACK-HEADED REED-WREN.)

Donacobius atricapillus, *Scl. Cat. A. B.* p. 16; *Scl. et Salv. Nomencl.* p. 5; *Sharpe, Cat. B.* vi. p. 364; *Burm. La-Plata Reise,* ii. p. 475. **Donacobius brasiliensis,** *d'Orb. Voy., Ois.* p. 213 (Corrientes).

Description.—Above dark chocolate-brown; cap black; wings black, with a large white patch on the inner primaries; tail black; lateral rectrices broadly ended with white; beneath ochreous buff; sides of breast and flanks with cross lines of dusky brown: total length 7·5 inches, wing 2·9, tail 3·7. *Female* similar.

Hab. Guiana, Amazonia, Brazil, Paraguay, and Northern Argentina.

The genus *Donacobius* contains two species somewhat intermediate between the Mock-birds and the large Wrens of the genus *Campylorhynchus.* The well-known Brazilian *D. atricapillus* extends through Paraguay, where Azara found it abundant, into Corrientes and the adjoining provinces of La Plata. It is met with among the reeds on lakes and streams.

---

12. TROGLODYTES FURVUS (Gm.).
(BROWN HOUSE-WREN.)

Troglodytes furvus, *Scl. et Salv. P. Z. S.* 1869, p. 158 (Conchitas); *iid. Nomencl.* p. 7; *Durnford, Ibis,* 1876, p. 157, 1877, p. 32 (Chupat), p. 167 (Buenos Ayres), 1878, p. 392 (Central Patagonia); *White, P. Z. S.* 1882, p. 593 (Buenos Ayres); *Döring, Exp. al Rio Negro, Zool.* p. 36 (Azul, R. Colorado, R. Negro). **Troglodytes platensis,** *Burm. La-Plata Reise,* ii. p. 476; *Barrows, Bull. Nutt. Orn. Cl.* viii. p. 86 (Concepcion). **Troglodytes musculus,** *Sharpe, Cat. B.* vi. p. 255.

Description.—Above brown; the tail-feathers and outer webs of wing-feathers pencilled with dark wavy lines; beneath very pale brown; bill and feet horn-colour; eye brown: total length 4·8 inches, wing 2·0, tail 1·7. *Female* similar.

Hab. South America.

The common Argentine Wren is to all English residents the "House-Wren," and is considered to be identical with the species familiar to them in their own country. It is a sprightly little bird, of a uniform brown colour and a cheerful melodious voice; a tireless hunter after small spiders and caterpillars in hedges, gardens, and outhouses, where it explores every dark hole and cranny, hopping briskly about with tail erect, and dropping frequent little curtsies; always prompt to scold an intruder with great emphasis; a great hater of cats.

It was my belief at one time that the Wren was one of the little birds a cat never could catch; but later on I discovered that this was a mistake. At my home on the pampas we once had a large yellow tom cat exceedingly dexterous in catching small birds; he did not, however, eat them himself, but used to bring them into the house for the other cats. Two or three times a day he would appear with a bird, which he would drop at the door, then utter a loud mew very well understood by the other cats, for they would all fly to the spot in great haste, and the first to arrive would get the bird. At one time I noticed that he brought in a Wren almost every day, and, curious to know how he managed to capture so clever a bird, I watched him. His method was to go out into the grounds frequented by Wrens, and seat himself conspicuously amongst the weeds or bushes; and then, after the first burst of alarm had subsided amongst the small birds, one or two Wrens would always take on themselves the task of dislodging him, or, at all events, of making his position a very uncomfortable one. The cat would sit perfectly motionless, apparently not noticing them at all, and by-and-by this stolid demeanour would have its effect, and one of the Wrens, growing bolder, would extend his dashing little incursions to within a few inches of pussy's demure face; then at last, swift as lightning, would come the stroke of a paw, and the little brown body would drop down with the merry brave little spirit gone from it.

The House-Wren is widely distributed in South America, from the tropical forests to the cold uplands of Patagonia, and, possessing a greater adaptiveness than most species, it inhabits every kind of country, moist or dry, and is as much at home on lofty mountains and stony places as in the everglades of the Plata, where it frequents the reed-beds and damp forests. About houses they are always to be found; and though the traveller on the desert pampas might easily imagine that there are no Wrens in the giant grasses, if he makes himself a lodge in this lonely region, a Wren will immediately appear to make its nest in his thatch and cheer him with its song.

Even in large towns they are common, and I always remember one flying into a church in Buenos Ayres one Sunday, and, during the whole sermon-time, pouring out its bright lyric strains from its perch high up somewhere in the ornamental wood-work of the roof.

The Wren sings all summer, and also on bright days in winter. The song is not unlike that of the English House-Wren, having the same gushing character, the notes being strong and clear, and uttered with rapidity and precision; but the Argentine bird has greater sweetness and more power.

In spring the male courts his mate with notes high and piercing as the squeals of a young mouse; these he repeats with great rapidity, fluttering his wings all the time like a moth, and at intervals breaking out into song.

The nest is made in a dark hole in a wall or tree, sometimes in the forsaken domed nest of some other bird; and where such sites are not to be found, in a dense thistle or thorn-bush, or in a large tussock of grass. I have also found nests in dry skulls of cows and horses, in an old boot, in the sleeve of an old coat left hanging on a fence, in a large-necked bottle, and in various other curious situations. The nest is built of sticks and lined with horse-hair or feathers, and the eggs are usually nine in number, of a pinkish ground-colour, thickly spotted with pale red.

---

13. TROGLODYTES AURICULARIS, Cab.

(EARED WREN.)

Troglodytes (Uropsila) auricularis, *Cab. Journ. f. Orn.* 1883, p. 105, t. ii. fig. 1.

Description.—In habit and size near the European Wren, *T. parvulus*, but peculiar for the blackish-brown hinder half of the ear-coverts and its broad white superciliaries. Upper surface and flanks brown; throat and middle of belly whitish, tinged with brownish yellow; wings and tail with fine black cross bands; crissum with broader black and white cross bands. (*Cabanis.*)

Hab. Tucuman.

This is a recent discovery of Herr Schulz in the Sierra of Tucuman. We have not yet met with specimens of it.

---

14. CISTOTHORUS PLATENSIS (Lath.).

(PLATAN MARSH-WREN.)

Cistothorus platensis, *Scl. et Salv. P. Z. S.* 1869, p. 158; *iid. Nomencl.* p. 7 *Durnford, Ibis,* 1877, p. 168 (Buenos Ayres); *Döring, Exp. al Rio Negro, Zool.* p. 37 (R. Sauce, R. Colorado, R. Negro); *Barrows, Bull. Nutt. Orn. Cl.*

viii. p. 87 (Carhué); *Sharpe, Cat. B.* vi. p. 244. **Cistothorus fasciolatus,** *Burm. La-Plata Reise*, ii. p. 476 (Mendoza).

Description.—Above pale sandy brown, variegated with black streaks; head darker brown, streaked with black; the hind neck paler brown, with narrow black streaks; wing-coverts brown; tail-feathers dark sandy brown, barred with blackish brown; under surface pale sandy buff: total length 4·3 inches, wing 1·85, tail 1·6. *Female* similar.

Hab. Argentina, Patagonia, and Falkland Islands.

This small Wren is rarely seen, being nowhere common, although widely distributed. It prefers open grounds covered with dense reeds and grasses, where it easily escapes observation. I have met with it near Buenos Ayres city; also on the desert pampas, in the tall pampas-grass. It is likewise met with along the Paraná river, and in Chili, Patagonia, and the Falkland Islands. In the last-named locality Darwin found it common, and says that it has there an extremely feeble flight, so that it may easily be run down and taken.

The Marsh-Wren has a sweet and delicate song, resembling that of the House-Wren (*Troglodytes furvus*), but much less powerful. It does not migrate; and on the pampas I have heard it singing with great animation when the pampas-grass, where it sat perched, was white with frozen dew. Probably its song, like that of *Troglodytes furvus*, varies in different districts; at all events, the pampas bird does not possess so fine a song as Azara ascribes to his "Todo Voz" in Paraguay, which is undoubtedly the same species.

Fam. V. MOTACILLIDÆ, or WAGTAILS.

The Wagtails and Pipits are closely-allied forms, and are usually referred to the same family of Oscines. The Wagtails are restricted to the Old World, although it has been recently ascertained that some of them occasionally occur as stragglers in the northern latitudes of America. Of the almost cosmopolitan Pipits about eight or nine species are sparingly distributed over the prairies and pampas of the New World. One of these is a common resident in the pampas of Argentina, and another (perhaps somewhat doubtful species) is occasionally met with.

---

15. ANTHUS CORRENDERA, Vieill.

(CACHILA PIPIT.)

Anthus correndera, *Scl. et Salv. Nomencl.* p. 8; *Hudson, P. Z. S.* 1873, p. 771 (Buenos Ayres); *Durnford, Ibis,* 1877, p. 32 (Chupat), p. 168 (Buenos Ayres), 1878, p. 392 (Central Patagonia); *Sclater, Ibis,* 1878, p. 362; *White, P. Z. S.* 1882, p. 594 (Buenos Ayres); *Döring, Exp. al Rio Negro, Zool.* p. 37 (Azul); *Barrows, Bull. Nutt. Orn. Cl.* viii. p. 87 (Concepcion, Entrerios); *Sharpe, Cat. B.* x. p. 610. **Anthus rufus,** *Burm. La-Plata Reise,* ii. p. 474 (Mendoza, Paraná); *Durnford, Ibis,* 1876, p. 158.

Description.—Above pale sandy buff, mottled with black centres to the feathers; wing- and tail-feathers dark brown, edged with buff, the penultimate tail-feather with a white tip, the outer tail-feather almost entirely white; neck and breast sandy buff, with large triangular black spots; flanks buff, streaked with black; abdomen and under tail-coverts isabelline; bill dusky grey; feet pink: total length 6·0 inches, wing 2·9, tail 2·3. *Female* similar.

Hab. Paraguay, Argentina, Patagonia, and Chili.

Azara's only reason for calling this bird *La Correndera* was because he thought it resembled a Tit-Lark known by that name in his own country, but of which he merely had a confused recollection. It is therefore to be regretted, I think, that *correndera* has been adopted as a specific name by naturalists instead of "Cachila," the vernacular name of the bird, familiar to every one in the Argentine country. Azara's Spanish bird was probably *Anthus pratensis,* which closely resembles *A. correndera* in general appearance, and has, moreover, as wide a range in the northern as the last-named species has in the southern hemisphere. In the volume on Birds in the 'Voyage of the Beagle,' it is said that a species of *Anthus* ranges further south than any other land-bird, being the only land-bird found on Georgia and South Orkney (lat. 61° S.).

In colour and language, possibly also in size, the Cachila is variable. It is a very common bird, widely and plentifully distributed over the pampas, found alike on marshy and dry grounds, but rare in the region of giant grasses. While abundant, it is also very evenly dispersed, each bird spending its life on a very circumscribed plot of earth. Those frequenting marshy or moist grounds are of a yellowish-cream colour, thickly mottled and striped with fuscous and black, and have two narrow parallel pure white marks on the back, very conspicuous when the bird is on the ground. The individuals frequenting high and dry grounds are much paler in hue, appearing almost grey, and do not show the white marks on the back. They also look larger than the birds on marshy lands; but this appearance is probably due to a looser plumage. The most strongly-marked pale and dark-plumaged variations may be found living within a few hundred yards of each other, showing how strictly each bird keeps to its own little "beat"; for this difference in coloration is, no doubt, due entirely to the amount of moisture in the ground they live on.

The Cachilas are resident, living in couples all the year round, the sexes being faithful. Several pairs frequent a small area, and sometimes they unite in a desultory flock; but these gatherings are not frequent. In the evening, at all seasons, immediately after the sun has set, the Cachilas all rise to a considerable height in the air and fly wildly about, chirping for a few minutes, after which they retire to roost.

When approached they frequently rise up several feet from the ground and flutter in the air, chirping sharply, with breast towards the intruder. This is a habit also found in Synallaxine species inhabiting the grassy plains. But, as a rule, the Cachilas are the tamest of feathered creatures, and usually creep reluctantly away on their little pink feet when approached. If the pedestrian is a stranger to their habits they easily delude him into attempting their capture with his hat, so little is their fear of man.

To sing, the Cachila mounts upwards almost vertically, making at intervals a fluttering pause, accompanied with a few hurried notes. When he has thus risen to a great height (but never beyond sight as Azara says) he begins the descent slowly, the wings inclining upwards; and, descending, he pours forth long impressive strains, each ending with a falling inflection or with two or three short throat-notes as the bird pauses fluttering in mid-air, and then renewed successively until, when the singer is within 3 or 4 feet of the earth, without alighting he reascends as before to continue the performance. It is a very charming melody, and heard always on the treeless plains when there is no other bird-music, with the exception of the trilling and grasshopper-like notes of a few Synallaxine species. But in character it is utterly unlike the song of the Sky-Lark with its boundless energy, hurry, and abandon; and yet it is impossible not to think of the Sky-Lark when describing the Cachila, which, in its manners, appearance, and in its habit of soaring to a great height when singing, seems so like a small copy of that bird.

The Cachila rears two broods in the year; the first is hatched about the middle of August, that is, one to three months before the laying-season of other Passerine species. By anticipating the breeding-season their early nests escape the evil of parasitical eggs; but, on the other hand, frosty nights and heavy rains are probably as fatal to as many early broods as the instinct of the *Molothrus bonariensis*, or Cow-bird, is to others at a later period.

The second brood is reared in December, the hottest month, and in that season a large proportion of their nests contain parasitical eggs.

The nest is placed in a slight hollow in the ground, under a tussock of grass, and is sometimes elaborately made and lined with horsehair and fine grass, and sometimes with a few materials loosely put together. During the solstitial heats I have frequently found nests with frail shades, built of sticks and grass, over them, the short withered grass affording an insufficient protection from the meridian sun. The eggs are four, elongated, with a dirty white and sometimes a dull bluish ground, thickly spotted with dusky brown and drab. In some eggs the spots are confluent, the whole shell being of a dull brownish-drab colour.

The manners of this species, where I have observed it, are always the same; it lives on the ground on open plains, where the herbage and grass is short, and never perches on trees. The song varies considerably in different districts.

---

16. ANTHUS FURCATUS, d'Orb. et Lafr.

(FORKED-TAIL PIPIT.)

Anthus furcatus, *d'Orb. Voy.* p. 227 (Patagonia); *Darwin, Zool. Voy. 'Beagle,'* iii. p. 85 (La Plata); *Sclater, Ibis,* 1878, p. 364; *Döring, Exp. al Rio Negro, Zool.* p. 37 (Azul, Carhué-pampas); *Sharpe, Cat. B.* x. p. 605.

Description.—Similar to *A. correndera*, but with a smaller bill, shorter and more curved hind claw, less spotted under surface, and different marking of the second outer rectrix, which has a clear and distinct white line along the inner side of the shaft: total length 6·0 inches, wing 3·2, tail 2·4.

Hab. Peru, Bolivia, and Argentina.

One of the Pipits procured at Conchitas belongs to this species, if distinct from the former. I think I recollect it as a resident on the pampas, closely resembling the Cachila in flight and language, but much shyer, and usually found concealed under Tulu grass on dry grounds.

Fam. VI. MNIOTILTIDÆ, OR WOOD-SINGERS.

The Mniotiltidæ, or Wood-singers, are a well-known and very characteristic family of the New World, where they occupy the position of our Sylviidæ. They number some 130 or 140 species, distributed all over America down to La Plata, but most abundant in the southern portions of North America, where the favourite and beautiful genus *Dendrœca*, with about 100 species, plays an important part. In Argentina only four species have as yet been met with.

17. PARULA PITIAYUMI (Vieill.).

(PITIAYUMI WOOD-SINGER.)

Parula pitiayumi, *Scl. et Salv. Nomencl.* p. 8; *Durnford, Ibis,* 1876, p. 158, 1877, p. 168 (Buenos Ayres); *Salv. Ibis,* 1880, p. 352 (Tucuman); *White, P. Z. S.* 1882, p. 594 (Catamarca, Misiones); *Barrows, Bull. Nutt. Orn. Cl.* viii. p. 87 (Concepcion, Entrerios); *Sharpe, Cat. B.* x. p. 259, pl. xi. fig. 1.
Sylvicola venusta, *Burm. La-Plata Reise,* ii. p. 473 (Paraná, Tucuman).

Description.—Above clear blue; mantle and upper back olive-yellow; central tail-feathers blue, all the others, also the quills, blackish; cheeks and under surface of body bright yellow; lower abdomen and under tail-coverts white; upper mandible black, lower yellow; eye brown: total length 4·0 inches, wing 2·05, tail 1·55. *Female* similar, but much paler in colour.

Hab. South America.

This is a southern representative of a small group of Wood-warblers, which is extensively diffused in the New World.

The upper plumage of this small bird is mostly cerulean-blue, the breast and belly yellow. Its Guarani name, according to Azara, is "*Pitiayume,*" which means little yellow-breast. I have never heard it sing or utter any note beyond a very feeble chirp as it hops about through the foliage in quest of small caterpillars. Its migration extends south to Buenos Ayres, where it is seen in woods and thickets in pairs or singly; but it is a rare bird, and I have been unable to find out anything about its nesting-habits.

---

18. GEOTHLYPIS VELATA (Vieill.).

(VEILED WOOD-SINGER.)

Geothlypis velata, *Scl. et Salv. Nomencl.* p. 9; *White, P. Z. S.* 1882, p. 594 (Salta); *Barrows, Bull. Nutt. Orn. Cl.* viii. p. 87 (Concepcion, Entrerios); *Sharpe, Cat. B.* x. p. 363, pl. ix. fig. 5.

Description.—Above yellowish green; the wing-coverts like the back; wing-feathers dusky brown, edged with olive-yellow; tail-feathers olive-green; crown of head to the occiput blue-grey; from the forehead a black mark extends to the eye and downward to the cheek; throat and under surface bright yellow; bill black; feet pale brown; eye brown: total length 5·6 inches, wing 2·4, tail 2·4. *Female* similar, but without the black on the face.

Hab. Brazil, Paraguay, and Northern Argentina.

This is again the only species of a North- and Central-American genus which ranges so far south as Buenos Ayres. It visits us in summer, and is found singly or in pairs in woods and large plantations. It feeds both on the ground and in trees, and, while gleaning amongst the leaves, frequently pauses to utter its loud cheerful song, composed of seven or eight clear notes uttered with rapidity and emphasis.

---

19. BASILEUTERUS AURICAPILLUS, Sw.

(GOLDEN-CROWNED WOOD-SINGER.)

Basileuterus auricapillus, *Sharpe, Cat. B.* x. p. 393. **Basileuterus vermivorus,** *Scl. P. Z. S.* 1865, p. 283; *Scl. et Salv. Nomencl.* p. 10; *White, P. Z. S.* 1882, p. 594 (Misiones).

Description.—Above olive-yellow, lightest on the rump and upper tail-coverts; tail-feathers ashy brown, with yellowish margins; quills dusky; crown of head light orange-chestnut; nape and hind neck pale ashy grey; on each side of the crown a broad black stripe extending from the bill to the hind neck, also a streak of ashy white above the eye; under surface bright yellow; axillaries and under wing-coverts white; bill and feet brownish: total length 4·5 inches, wing 2·2, tail 2·05. *Female* similar.

Hab. South America.

This species, which is widely spread over the northern portion of South America, was found in Paraguay by Azara, and in the province of Misiones by White.

---

20. SETOPHAGA BRUNNEICEPS, d'Orb. et Lafr.

(BROWN-CAPPED WOOD-SINGER.)

Setophaga brunneiceps, *Burm. La-Plata Reise,* ii. p. 473 (Tucuman); *Scl. et Salv. Nomencl.* p. 11; *White, P. Z. S.* 1882, p. 595 (Catamarca); *Sharpe, Cat. B.* x. p. 428.

Description.—General colour dull olive-yellowish; wing-coverts dark slaty grey; quills blackish; upper tail-coverts slaty grey, washed with olive; tail-feathers blackish—the outer pair white, the next pair white edged with black on the outer web, the third pair with a large white mark at the end; crown of head deep chestnut; neck slaty grey; under surface of body bright yellow; under tail- and wing-coverts white: total length 5·0 inches, wing 2·45, tail 2·4. *Female* similar.

Hab. Bolivia and Northern Argentina.

White obtained a pair of these birds on the Sierra of Totoral, Catamarca, in July 1880. He describes them as quick in their movements and difficult to shoot. The species was originally discovered by d'Orbigny in Bolivia.

Fam. VII. VIREONIDÆ, or GREENLETS.

Three of the groups of the peculiar American family of Greenlets, allied to our Shrikes, have representatives within the Argentine Republic. The genera *Vireo* and *Hylophilus* both extend, each in the shape of one of its South-Brazilian members, into the woodlands of the Paraná; while *Cyclorhis*, another genus also widely spread over South and Central America, has two representatives within our area. One of these latter is well known in the neighbourhood of Buenos Ayres; the other is only found in the extreme north of the Republic.

---

21. VIREOSYLVIA CHIVI (Vieill.).

(CHIVI GREENLET.)

Contramaestre gaviero, *Azara, Apunt.* ii. p. 34. **Sylvia chivi**, *Vieill. N. D.* xi. p. 174. **Vireosylvia chivi**, *Baird, Rev. A. B.* p. 337; *Scl. et Salv. P. Z. S.* 1869, p. 160 (Buenos Ayres); *Berl. et Jher. Zeitschr. ges. Orn.* 1885, p. 115. **Vireo chivi**, *Gadow, Cat. B.* viii. p. 295.

Hab. South America, from Colombia down to Buenos Ayres.

A single specimen of this Greenlet was found in a collection made by Mr. Haslehurst near Buenos Ayres. As the species occurs in Rio Grande do Sul (*Berlepsch*) and Paraguay (*Azara*), its occasional appearance in Eastern Argentina is quite probable. Whether the bird is really distinct from the widespread *Vireo olivaceus* seems to be a question which is not yet finally settled.

Azara, describing this species, says it is one of the commonest in deep woods, where it moves about among the terminal twigs, without ever rising to the tops of the trees or flying down to the brush or the ground. It is active, and extremely restless in manner; and in searching after and taking the small insects and spiders on which it lives it climbs about the twigs, assuming every position, and frequently suspending itself, head downwards, by its feet. It has a full pleasing voice of considerable power, heard incessantly in the woods, particularly in the love-season. The nest is a slender beautiful structure, even surpassing that of the Humming-birds, constructed of thin dry leaves outside, smoothly attached to the rest with spiders' webs, while the inside is formed of fine fibres and cotton.

---

22. HYLOPHILUS PŒCILOTIS, Max.

(BROWN-HEADED WOOD-BIRD.)

Hylophilus pœcilotis, *Scl. et Salv. Nomencl.* p. 12; *Scl. Ibis,* 1881, p. 300; *White, P. Z. S.* 1882, p. 595 (Misiones); *Gadow, Cat. B.* viii. p. 308.

Description.—Crown of head and nape rich rufous-brown; all the rest of the upper parts, including the entire tail and wing-coverts and the outer webs of the remiges, rich olive-green; cheeks and upper throat whitish; ear-coverts blackish, with white central streaks; under wing-coverts, axillaries, inner margin of remiges, and under tail-coverts lemon-yellow; rest of underparts dull yellow, washed with olive on the breast and flanks, and inclining to pale ochreous on the abdomen; bill dark flesh-colour; feet hazel: total length 4·5 inches, wing 2·2, tail 2·2. *Female* similar.

Hab. Southern Brazil, Paraguay, and Northern Argentina.

This South-Brazilian species was met with by White near San Javier, in the province of Misiones, in June 1881.

---

23. CYCLORHIS OCHROCEPHALA.

(OCHRE-HEADED GREENLET-SHRIKE.)

[Plate III. Fig. I.]

Cyclorhis viridis, *Burm. La-Plata Reise,* ii. p. 472; *Scl. et Salv. Nomencl.* p. 13; *Durnford, Ibis,* 1878, p. 58 (Punta Lara); *White, P. Z. S.* 1882, p. 595 (Buenos Ayres); *Gadow, Cat. B.* viii. p. 318. **Cyclorhis ochrocephala,** *Tsch. Arch. f. Nat.* 1845, pt. i. p. 362; *Berl. et Jher. Zeitschr. ges. Orn.* 1885, p. 116.

Description.—Above olive-green; cap brownish ochraceous, more or less rufescent; front and superciliaries chestnut-red; sides of head clear grey, beneath pale buff; breast and flanks yellow; throat greyish white; bill reddish grey, feet grey; eye reddish: total length 7·0 inches, wing 3·4, tail 2·8. *Female* similar.

Hab. Middle districts of Argentina.

This species is not uncommon in the woods along the shores of the Plata, and may be easily known to any person penetrating them by its loud "cheerful soliloquy," for that phrase of Mr. Barrows, the North-American writer on birds, well describes the artless, light-hearted song which it utters at intervals while it roams about in the deep foliage, and which reminds one of the careless whistling of a boy, whistling merely to express his gaiety, but without having any particular tune in his mind. It is migratory, and extends its range south of Buenos Ayres.

---

24. CYCLORHIS ALTIROSTRIS, Salvin.

(DEEP-BILLED GREENLET-SHRIKE.)

[PLATE III. FIG. 2.]

Cyclorhis altirostris, *Salv. Ibis,* 1880, p. 352; *Gadow, Cat. B.* viii. p. 319.
Cyclorhis viridis, *Barrows, Bull. Nutt. Orn. Cl.* viii. p. 88 (Concepcion, Entrerios); *Berl. et Jher. Zeitschr. ges. Orn.* 1885, p. 116.

Description.—Above olive-green; head more or less rufescent; front and superciliaries chestnut-red; sides of head grey; beneath pale ochraceous; breast and sides yellow; throat greyish; bill short and thick, pale reddish, with a black blotch at the base of the lower mandible; feet reddish: total length 6·5 inches, wing 3·3, tail 3·0. *Female* similar.

Hab. Paraguay and Northern Argentina.

This species was met with by Durnford near Salta in June 1878. He describes the iris as "light rufous; upper mandible dark slate, under mandible, legs, and feet pale slate."

Mr. Salvin founded his *C. altirostris* upon Durnford's specimens, but Graf v. Berlepsch is of opinion that this species is the "*Habia verde*" of Azara, and should consequently bear the name "*viridis*" of Vieillot. This is perhaps correct, but at the same time it would only make fresh confusion to transfer to this species the name hitherto usually applied to the preceding bird. We prefer, consequently, to let it stand under Mr. Salvin's name "*altirostris.*"

Fam. VIII. HIRUNDINIDÆ, OR SWALLOWS.

The cosmopolitan family of Swallows, of which about eighty species are known, is well developed in the New World, where some thirty representatives occur in various parts. In Argentina the occurrence of eight Swallows has been recorded. Three of them belong to the group of Purple Martins (*Progne*), which is restricted to the New World, and of the remainder four are members of genera entirely restricted to the Neotropical Region. The genus *Petrochelidon,* of which one species is met with in La Plata, has alone representatives in the Eastern Hemisphere.

---

25. PROGNE FURCATA, Baird*.

(PURPLE MARTIN.)

Progne purpurea, *Hudson, P. Z. S.* 1872, p. 548 (Patagonia); *Scl. P. Z. S.* 1872, p. 605; *Scl. et Salv. Nomencl.* p. 14; *Durnford, Ibis,* 1877, p. 32 (Chupat),

* [I here follow Mr. Sharpe in considering the Purple Martin of Argentina and Patagonia separable from the closely allied northern species of North America. But I am altogether sceptical about its occurrence in Chili, which Dr. Philippi expressly daries (P. Z. S. 1868 p. 534) and for which I know of no good authority.—P. L. S.]

Fig. 1. CYCLORHIS OCHROCEPHALA.
2. ALTIROSTRIS.

1878, p. 392 (Central Patagonia); *White, P. Z. S.* 1882, p. 595 (Buenos Ayres). **Progne elegans,** *Barrows, Bull. Nutt. Orn. Cl.* viii. p. 89 (Bahia Blanca). **Progne furcata,** *Baird, Rev. A. B.* p. 278; *Sharpe, Cat. A. B.* x. p. 175.

Description.—Uniform deep purple-blue; on each side of the back a small concealed tuft of white feathers; tail-feathers black, washed with blue: total length 7·7 inches, wing 5·55, tail 3·3. *Female*: upper parts dull purple; head, neck, and lower parts blackish brown.

Hab. Argentina and Patagonia.

The Purple Martin is occasionally seen in the eastern provinces of La Plata when migrating, but has not been found nesting anywhere so far north as Buenos Ayres. I met with it breeding at Bahia Blanca on the Atlantic coast, and on the Rio Negro, where it is very common. It arrives in Patagonia late in September, and leaves before the middle of February. On the 14th of that month I saw one flock flying north, but it was the last. It breeds in holes under the eaves of houses or in walls, and its nest is like that of *P. chalybea*; but many also breed in holes in the steep banks of the Rio Negro. They do not, however, excavate holes for themselves, but take possession of natural crevices and old forsaken burrows of the Burrowing Parrot (*Conurus patachonicus*). In size, flight, manners, and appearance the Purple Martin closely resembles the following species, the only difference being in the dark plumage of the under surface. The language of the two birds is also identical; the loud excited scream when the nest is approached, the various other notes when the birds sweep about in the air, and the agreeably modulated and leisurely-uttered song are all possessed by the two species without the slightest difference in strength or intonation. This circumstance appears very remarkable to me, because, though two species do sometimes possess a few notes alike, the greater part of their language is generally different; also because birds of the same species in different localities vary more in language than in any other particular. This last observation, however, applies more to resident than to migratory species.

---

26. PROGNE CHALYBEA (Gm.).

(DOMESTIC MARTIN.)

Progne chalybea, *Scl. P. Z. S.* 1872, p. 606 (Buenos Ayres); *Scl. et Salv. Nomencl.* p. 14; *White, P. Z. S.* 1882, p. 595 (Buenos Ayres); *Barrows, Bull. Nutt. Orn. Cl.* viii. p. 88 (Concepcion); *Sharpe, Cat. B.* x. p. 178. **Progne domestica,** *Burm. La-Plata Reise,* ii. p. 477; *Sharpe, Cat. B.* x. p. 177.

Description.—Entire upper parts deep purple-blue; lesser and median wing-coverts the same; wing- and tail-feathers black, glossed with steel-blue; throat, fore neck, and chest ash-colour; breast, abdomen, and under tail-coverts pure white; bill and feet black: total length 8 inches, wing 5·7, tail 3·2. *Female* similar.

Hab. Central and South America.

This species, distinguished from the Purple Martin by its white underparts, ranges from Mexico to Buenos Ayres, the extreme limit of its range being about 250 miles south of that city. It was well called *Golondrina domestica* by Azara, being preeminently domestic in its habits. It never breeds in banks as the Purple Martin often does, or in the domed nests of other birds in trees, a situation always resorted to by the Tree-Martin, and occasionally by the Common Swallow; but is so accustomed to the companionship of man as to make its home in populous towns as well as in country-houses. It arrives in Buenos Ayres about the middle of September, and apparently resorts to the same breeding-place every year. A hole under the eaves is usually selected, and the nest is roughly built of dry grass, hair, feathers, and other materials. When the entrance to its breeding-hole is too large, it partially closes it up with mud mixed with straw; if there be two entrances it stops up one altogether. The bird does not often require to use mud in building; it is the only one of our Swallows that uses such a material at all. The eggs are white, long, pointed, and five in number.

In the season of courtship this Martin is a noisy pugnacious bird, and always, when quitting its nest, utters an exceedingly loud startling cry several times repeated. It also has a song, uttered both when resting and on the wing, composed of several agreeably modulated notes, and in that thick rolling intonation peculiar to our Swallows. This song does not sound loud when near, yet it can be distinctly heard when the bird appears but a speck in the distance. I may here remark that, with the exception of the *Petrochelidon pyrrhonota*, which possesses a sharp squeaky voice, like the Swallows of Europe, all our Hirundines have soft voices: their usual twittering when they are circling about resembles somewhat the chirping of the English House-Sparrow in tone, but besides these notes they possess a song more pleasing to the ear.

Before leaving in February these birds congregate in parties of from twenty to four or five hundred, usually on the broad leafy top of an old ombú tree.

---

27. PROGNE TAPERA (Linn.).

(TREE-MARTIN.)

Progne tapera, *Scl. P. Z. S.* 1872, p. 606 (Buenos Ayres); *Scl. et Salv. Nomencl.* p. 14; *Durnford, Ibis*, 1877, p. 168 (Buenos Ayres); *White, P. Z. S.* 1882, p. 595 (Catamarca); *Barrows, Bull. Nutt. Orn. Cl.* viii. p. 89 (Concepcion); *Sharpe, Cat. B.* x. p. 180. **Cotyle tapera**, *Burm. La-Plata Reise*, ii. p. 477.

Description.—Entire upper parts dull brown; tail-feathers blackish brown; throat ashy white; fore neck and chest ashy brown in the centre and dark brown on the sides, as are also the flanks; abdomen white; bill and feet horn-colour: total length 7 inches, wing 5·5, tail 2·6. *Female* similar.

Hab. Brazil, Paraguay, Argentina.

The Tree-Martin is more slender and has a greater extent of wing than the Purple Martins; and, instead of the beautiful dark purple (their prevailing colour), its entire upper plumage is dusky brown, the under surface white. But if these differences of structure and hue merely serve to show that it is not a very near relative of the other species, those exhibited in its habits remove it very far indeed from them.

The Tree-Martin is a very garrulous bird, and no sooner arrives early in September, than we are apprised of the circumstance by the notes which the male and female incessantly sing in concert, fluttering and waving their wings the while, and seeming quite beside themselves with joy at their safe arrival; for invariably they arrive already mated. Their language is more varied, the intonation bolder and freer than that of our other Swallows. The length of the notes can be varied at pleasure; some are almost harsh, others silvery or liquid, as of trickling drops of water; they all have a glad sound; and many have that peculiar character of some bird-notes of shaping themselves into words.

This Martin is never seen to alight on the ground or on the roofs of houses, but solely on trees; and when engaged in collecting materials for its nest, it sweeps down and snatches up a feather or straw without touching the surface. It breeds only in the clay-ovens of the Oven-bird (*Furnarius rufus*). I, at least, have never seen them breed in any other situation after observing them for a great many summers. An extraordinary habit! for, many as are the species that possess the parasitical tendency of breeding in other birds' nests, none of them confine themselves to the nest of a single species excepting the bird I am describing. It must, however, be understood that my knowledge of this bird has been acquired in Buenos Ayres, where I have observed it; and as this Martin possesses a wider range in South America than the Oven-birds, it is more than probable that in other districts it builds in different situations.

On arriving in spring each pair takes up its position on some tree, and usually on a particular branch; a dead branch extending beyond the foliage is a favourite perch. Here they spend much of their time, never appearing to remain long absent from it, and often, when singing their notes together, fluttering about it with a tremulous uncertain flight, like that of a hovering butterfly. About three weeks after first arriving they begin to make advances towards the Oven-bird's nest that stands on the nearest post or tree; and if it be still occupied by the rightful owners, after much time has been spent in sporting about and reconnoitring it, a feud begins which is often exceedingly violent and protracted for many days.

In seasons favourable to them the Oven-birds build in autumn and winter, and breed early in spring; so that their broods are out of their clay-houses by the end of October or earlier; when this happens, the Swallow that breeds in November quietly takes possession of the forsaken fortress. But accidents will happen, even to the wonderful fabric of the Oven-bird. It is sometimes destroyed and must be rebuilt; or its completion has perhaps been retarded for months by drought, or by the poor condition of the birds in severe weather; or the first brood has perhaps perished, destroyed by an opossum or other enemy. November, and even December, may thus arrive before some pairs have hatched their eggs; and it is these unfortunate late breeders that suffer from the violence of the marauding Swallows. I have often witnessed the wars of these birds with the deepest interest; and in many ovens that I have opened I have found the eggs of the Oven-birds buried under the nests of the Swallows. After the Swallows have taken up a position near the coveted oven, they occasionally fly towards and hover about it, returning again to their stand. By-and-by, instead of returning as at first, they take to alighting at the entrance of the coveted home; this is a sort of declaration of war, and marks the beginning of hostilities. The Oven-birds, full of alarm and anger, rush upon and repel them as often as they approach; they retire before this furious onset, but not discomfited, and only warbling out their gay seemingly derisive notes in answer to the outrageous indignant screams of their enemies. Soon they return; the scene is repeated; and this desultory skirmishing is often continued for many days.

But at length the lawless invaders, grown bolder, and familiar with his strength and resources, will no longer fly from the master of the house; desperate struggles now frequently take place at the entrance, the birds again and again dropping to the ground clutched fiercely together, and again hurrying up only to resume the combat. Victory at last declares itself for the aggressors, and they busy themselves carrying in materials for their nest, screaming their jubilant notes all the time as if in token of triumph. The brave and industrious Oven-birds, dispossessed of their home, retire to spend their childless summer together, for the male and female never separate; and when the autumn rains have supplied them with wet clay, and the sense of defeat is worn off, they cheerfully begin their building-operations afresh.

This is not, however, the invariable result of the conflict. To the superior swiftness of the Martin the Oven-bird opposes greater strength, and, it might be added, a greater degree of zeal and fury than can animate its adversary. The contest is thus nearly an equal one; and the Oven-bird, particularly when its young are already hatched, is often able to maintain its own. But the Martins never suffer defeat; for, when unable to take the citadel by storm, they fall back on their dribbling system of warfare, which they keep up till the young birds leave the nest, when they take possession before it has grown cold.

The Martin makes its own nest chiefly of large feathers, and lays four eggs, long, pointed, and pure white.

It will be remarked that in all its habits above mentioned this bird differs widely from the two preceding species. It also differs greatly from them in its manner of flight. The Purple Martins move with surprising grace and celerity, the wings extended to their utmost; they also love to sail in circles high up in the air, or about the summits of tall trees, and particularly during a high wind. At such times several individuals are usually seen together, and all seem striving to outvie each other in the beauty of their evolutions.

The Tree-Martin is never seen to soar about in circles; and though when hawking after flies and moths it sweeps the surface of the grass with amazing swiftness, at other times it has a flight strangely slow and of a fashion peculiar to itself: the long wings are depressed as much as those of a Wild Duck when dropping on to the water, and are constantly agitated with tremulous flutterings, short and rapid as those of a butterfly.

Neither is this bird gregarious like all its congeners, though occasionally an individual associates for a while with Swallows of another species; but this only when they are resting on fences or trees, for as soon as they take flight it leaves them. Once or twice, when for some mysterious cause the autumnal migration has been delayed long past its usual time, I have seen them unite in small flocks; but this is very rare. As a rule they have no meetings preparatory to migration, but skim about the fields and open plains in un-Swallow-like solitude, and suddenly disappear without having warned us of their intended departure.

28. PETROCHELIDON PYRRHONOTA (Vieill.).

(RED-BACKED ROCK-MARTIN.)

Petrochelidon pyrrhonota, *Scl. et Salv. Nomencl.* p. 14; *Durnford, Ibis,* 1877, p. 169; *White, P. Z. S.* 1882, p. 595 (Buenos Ayres); *Sharpe, Cat. B.* x. p. 193. **Cotyle pyrrhonota,** *Burm. La-Plata Reise,* ii. p. 477 (Tucuman).

Description.—Above glossy dark steel-blue; lower back and rump cinnamon-rufous; the upper tail-coverts brown, with grey margins; wings black; tail black, with greenish gloss; crown steel-blue; forehead sandy buff; cheeks and sides of face chestnut, spreading to the sides of the hind neck; chin chestnut; the lower throat steel-blue; fore neck, chest, and sides of body and flanks light ashy brown; centre of breast and abdomen white, tinged with brown; under tail-coverts, also under wing-coverts and axillaries, ashy brown: total length 5·3 inches, wing 4·35, tail 2·05. *Female* similar.

Hab. South America.

This species does not breed in Buenos Ayres, and is only seen there in spring, flying south or south-west, and again in much larger numbers on its return journey in autumn. On the Rio Negro, in Patagonia, I did not meet with it, and suppose its summer resort must be south of that locality; and, judging from the immense numbers visible in some seasons, I should think that they must, in their breeding-place in Patagonia, occupy a very extensive area. They do not seem to be as regular in their movements as other Swallows here; some years I have observed them passing singly or in small parties during the entire hot season: usually they begin to appear, flying north, in February; but in some years not until after the middle of March. They are not seen passing with a rapid flight in close flocks, but straggle about, hawking after flies: first one bird passing, then two or three, and a minute or two later half a dozen, and so on for a greater part of the day. So long as the weather continues warm they journey in this leisurely manner; but I have known them to continue passing till April, after all the summer migrants had left us, and these late birds flew by with great speed in small close flocks, directly north, as if their flight had been guided by the magnetic needle.

While flying this species continually utters sharp twitterings and grinding and squealing notes of various lengths.

---

29. TACHYCINETA LEUCORRHOA (Vieill.).

(WHITE-RUMPED SWALLOW.)

Hirundo leucorrhoa, *Hudson, P. Z. S.* 1872, pp. 606, 845 (Buenos Ayres); *Scl. et Salv. Nomencl.* p. 14; *Durnford, Ibis,* 1877, p. 32 (Chupat), 1878, p. 392 (Central Patagonia); *White, P. Z. S.* 1882, p. 596 (Corrientes); *Barrows,*

Bull. Nutt. Orn. Cl. viii. p. 89 (Concepcion). **Cotyle leucorrhoea,** *Burm. La-Plata Reise*, ii. p. 478 (Paraná). **Tachycineta leucorrhous,** *Sharpe, Cat. B.* x. p. 114.

Description.—Above glossy dark green; rump white; quills black, washed with green; upper tail-coverts dark green; tail-feathers black with greenish gloss; base of forehead white, extending a little backward over the lores; cheeks and whole under surface white; flanks and sides washed with smoky brown; axillaries and under tail-coverts pale smoky brown; bill and feet black: total length 5·5 inches, wing 4·45, tail 2·0. *Female* similar.

Hab. Brazil, Paraguay, and Argentina.

This is the most abundant and best known of our Swallows; a pretty bird in its glossy coat of deep green, and rump and under surface snowy white; exceedingly restless in its disposition, quick and graceful in its motions; social, quarrelsome, garrulous, with a not unmusical song, beginning with long, soft, tremulous notes, followed by others shorter and more hurried, and sinking to a murmur. They are the last of all our migrants to leave us in autumn, and invariably reappear in small numbers about the houses on every warm day in winter. Probably many individuals in Buenos Ayres remain through the winter in sheltered situations, to scatter over the surrounding country whenever there comes a warm bright day. I once saw three together, skimming over the plains, on one of the coldest days I ever experienced on the pampas, the thermometer having stood at 29° Fahr. that morning.

Further south their migration is more strict; and on the Rio Negro, in Patagonia, from March to August I did not meet with a single individual. In Buenos Ayres the autumnal migration of the Hirundines begins about the middle of February, and from that date vast numbers of this Swallow are seen travelling north, and, in some seasons, they continue passing for over a month. One autumn, in April, several days after the Swallows had all disappeared, flocks of the Common Swallow began again to appear flying north, and for ten days afterwards they continued to pass in large numbers. They would stoop to dip themselves in a pool where I observed them, and then alight on the reeds and bushes to rest, and appeared quite tired with their journey, rising reluctantly when approached, and some allowing me to stand almost within arm's length of them without stirring. I had never before observed any later or supplementary migration like this; for, as a rule, the causes which in some years delay the departure of birds seems to affect them all alike. Possibly these late birds come from some remote district, where exceptionally cold weather had retarded breeding-operations.

The White-rumped Swallow sometimes lays in a tree, in the large nest, previously abandoned, of the Leñatero (*Anumbius acuticaudatus*). Its favourite site is, however, a hole in a wall, sheltered by the overhanging tiles or thatch; for though it does not go much into towns, as Azara has remarked, it is very domestic, and there is not a house on the pampas, however humble it be, but some of these birds are about it, sportively skimming above the roof, or curiously peering under the eaves, and incessantly uttering their gurgling happy notes.

For a period of a month to six weeks before building begins they seem to be holding an incessant dispute, reminding one in their scolding tones of a colony of contentious English House-Sparrows, only the Swallow has a softer, more varied voice, and frequently, even when hotly quarrelling, he pauses to warble out his pretty little song, with its sound like running water. However many eligible chinks and holes there may be, the contention is always just as great amongst them, and is doubtless referable to opposing claims to the best places. The excited twittering, the incessant striving of two birds to alight on the same square inch of wall, the perpetual chases they lead each other round and round the house, always ending exactly where they began, tell of clashing interests and of great unreasonableness on the part of some amongst them. By-and-by the quarrel assumes a more serious aspect; friends and neighbours have apparently intervened in vain; all the arguments of which Swallows are capable have been exhausted, and, a compromise of claims being more impossible than ever, fighting begins. Most vindictively do the little things clutch each other and fall to the earth twenty times an hour, where they often remain struggling for a long time, heedless of the screams of alarm their fellows set up above them; for often, while they thus lie on the ground punishing each other, they fall an easy prey to some wily pussy who has made herself acquainted with their habits.

When these feuds are finally settled, they address themselves diligently to the great work and build a rather big nest. They are not neat or skilful workers, but merely stuff a great quantity of straw and other light materials into the breeding-hole, and line the nest with feathers and horsehair. On this soft but disorderly bed the female lays from five to seven pure white eggs.

All those species that are liable at any time to become the victims of raptorial birds are very much beholden to this Swallow, as he is the most vigilant sentinel they possess. When the hurrying Falcon is still far off, and the other birds unsuspicious of his approach, the Swallows suddenly rush up into the sky with a wild rapid flight to announce the evil tidings with distracted screams. The alarm spreads swift as light through the feathered tribes, which, on all sides, are in terrified commotion, crouching in the grass, plunging into thickets, or mounting upwards to escape by flight. I have often wondered at this, since this swift-winged and quick-doubling little bird is the least likely to fall a prey himself.

They possess another habit very grateful to the mind of every early riser. At the first indication of dawn, and before any other wild bird has broken the profound silence of night, multitudes of this Swallow, as if at the signal of a leader, begin their singing and twittering, at the same time mounting upwards into the quiet dusky sky. Their notes at this hour differ from the hurried twittering uttered during the day, being softer and more prolonged, and, sounding far up in the sky from so many throats, the concert has a very charming effect, and seems in harmony with the shadowy morning twilight.

---

30. ATTICORA CYANOLEUCA (Vieill.).

(BANK-SWALLOW.)

Atticora cyanoleuca, *Burm. La-Plata Reise,* ii. p. 479; *Scl. et Salv. Nomencl.* p. 14; *Hudson, P. Z. S.* 1872, p. 844 (Buenos Ayres); *Durnford, Ibis,* 1876, p. 158 (Buenos Ayres), 1877, p. 32 (Chupat), p. 170 (Buenos Ayres), 1878, p. 392 (Central Patagonia); *White, P. Z. S.* 1882, p. 596 (Catamarca); *Barrows, Bull. Nutt. Orn. Cl.* viii. p. 90 (Concepcion, Bahia Blanca); *Sharpe, Cat. B.* x. p. 186.

Description.—Above dark glossy blue; quills and tail-feathers black; cheeks and under surface of body pure white, the sides of the neck blue, descending in a half-crescent on the sides of the chest; sides of body and flanks brown; under tail-coverts black; bill and feet black: total length 4·7 inches, wing 4·05, tail 2·2. *Female* similar.

Hab. Central and South America.

This diminutive dark-plumaged species is the smallest of our Hirundines. In Buenos Ayres they appear early in September, arriving before the Martins, but preceded by the Common Swallow. They are bank-birds, breeding in forsaken holes and burrows, for they never bore into the earth themselves, and are consequently not much seen about the habitations of man. They sometimes find their breeding-holes in the banks of streams, or, in cultivated districts, in the sides of ditches, and even down in wells. But if in such sites alone fit receptacles for their eggs were met with, the species, instead of one of the commonest, would be rare indeed with us; for on the level pampas most of the water-courses have marshy borders, or at most but low and gently sloping banks. But the burrowing habits of two other animals—the Vizcacha (*Lagostomus trichodactylus*), the common large rodent of the pampas, and the curious little bird called Minera (*Geositta cunicularia*) —have everywhere afforded the Swallows abundance of breeding-places on the plains, even where there are no streams or other irregularities in the smooth surface of the earth.

The Minera bores its hole in the sides of the Vizcacha's great burrow, and in this burrow within a burrow the Swallow lays its eggs and rears its young, and is the guest of the Vizcacha, and as much dependent on it as the House-Wren and the Domestic Swallow on man; so that in spring, when this species returns to the plains, it is in the villages of the Vizcachas that we see them. There they live and spend the day, sporting about the burrows, just as the Common Swallow does about our houses; and to a stranger on the pampas one of these villages, with its incongruous bird and mammalian inhabitants, must seem a very curious sight in the evening. Before sunset the old male Vizcachas come forth to sit gravely at the mouths of their great burrows. One or two couples of Mineras, their little brown bird-tenants, are always seen running about on the bare ground round the holes, resting at intervals with their tails slowly moving up and down, and occasionally trilling-out their shrill laughter-like cry. Often a pair of Burrowing-Owls also live in the village, occupying one of the lesser disused burrows; and round them all flit half a dozen little Swallows, like twilight-moths with long black wings. It is never quite a happy family, however, for the Owls always hiss and snap at a Vizcacha if he comes too near; while the little Swallows never become reconciled to the Owls, but perpetually flutter about them, protesting against their presence with long complaining notes.

The nest, made of dry grass lined with feathers, is placed at the extremity of the long, straight, cylindrical burrow, and contains five or six white pointed eggs. I have never seen these Swallows fighting with the Minera to obtain possession of the burrows, for this industrious little bird makes itself a fresh one every spring, so that there are always houses enough for the Swallows. After the young have flown, they sit huddled together on a weed or thistle-top, and the parents continue to feed them for many days.

As in size and brightness of plumage, so in language is the Bank-Swallow inferior to other species, its only song being a single, weak, trilling note, much prolonged, which the bird repeats with great frequency when on the wing. Its voice has ever a mournful, monotonous sound, and even when it is greatly excited and alarmed, as at the approach of a fox or hawk, its notes are neither loud nor shrill. When flying they glide along close to the earth, and frequently alight on the ground to rest, which is contrary to the custom of other Swallows. Like other species of this family, they possess the habit of gliding to and fro before a traveller's horse, to catch the small twilight-moths driven up from the grass. A person riding on the pampas usually has a number of Swallows flying round him, and I have often thought that more than a hundred were before my horse at one time; but, from the rapidity of their motions, it is impossible to count them. I have frequently noticed individuals of the four most common species following me together; but after sunset, and when the other species have long forsaken the open grassy plain for the shelter of trees and houses, the diminutive Bank-Swallow continues to keep the traveller company. At such a time, as they glide about in the dusk of evening, conversing together in low tremulous tones, they have a peculiarly sorrowful appearance, seeming like homeless little wanderers over the great level plains.

When the season of migration approaches they begin to congregate in parties not very large, though sometimes as many as one or two hundred individuals are seen together; these companies spend much of their time perched close together on weeds, low trees, fences, or other slightly elevated situations, and pay little heed to a person approaching, but seem preoccupied or preyed upon by some trouble that has no visible cause.

The time immediately preceding the departure of the Swallows is indeed a season of very deep interest to the observer of nature. The birds in many cases seem to forget the attachment of the sexes and their songs and aerial recreations; they already begin to feel the premonitions of that marvellous instinct that urges them hence: not yet an irresistible impulse, it is a vague sense of disquiet; but its influence is manifest in their language and gestures, their wild manner of flight, and their listless intervals.

The little Bank-Swallow disappears immediately after the Martins. Many stragglers continue to be seen after the departure of the main body; but before the middle of March not one remains, the migration of this species being very regular.

---

31. ATTICORA FUCATA (Temm.).

(BROWN MARTIN.)

Cotyle fucata, *Burm. La-Plata Reise,* ii. p. 478 (Mendoza); *Scl. et Salv. Nomencl.* p. 14; *White, P. Z. S.* 1882, p. 596 (Corrientes), 1883, p. 37 (Cordova).
Atticora fucata, *Sharpe, Cat. B.* x. p. 188.

Description.—Above brown; primary-coverts and quills blackish brown; tail-feathers dark brown; crown of head deep rufous, becoming clearer on the nape; cheeks, throat, and breast pale tawny; sides of body brown, tinged with rufous; centre of breast, abdomen, and under tail-coverts white; thighs, under wing-coverts, and axillaries brown: total length 4·6 inches, wing 4·15, tail 2·0. *Female* similar.

Hab. Guiana, Brazil, and Northern Argentina.

This Swallow is common near Mendoza, according to Prof. Burmeister. White obtained it in May 1881 at Santo Tomé, Corrientes, and in 1882 at Cosquin near Cordova. At Cosquin the first individual was seen on July 20th, but towards the end of August large numbers were observed, mostly skimming over the river.

---

32. STELGIDOPTERYX RUFICOLLIS (Vieill.).

(RED-NECKED SWALLOW.)

Stelgidopteryx ruficollis, *Sharpe, Cat. B.* x. p. 208. **Cotile ruficollis,** *Barrows, Bull. Nutt. Orn. Cl.* viii. p. 90.

Description.—Above brown, head darker, rump paler; wings and tail blackish brown, coverts edged with pale brown, external secondaries margined with whitish; beneath pale ashy brown; throat rufous; middle of belly pale yellowish; crissum white, tipped with black: total length 5·0 inches, wing 4·3, tail 2·0. *Female* similar.

Hab. South America.

Mr. Barrows tells us that this species is abundant in Entrerios throughout the summer, arriving from the north early in August, and is said to nest in holes in banks.

Fam. IX. TANAGRIDÆ, OR TANAGERS.

The brilliant family of Tanagers, one of the most characteristic groups of the American avifauna, contains altogether nearly 400 species, of which the greater number are restricted to the forest-clad districts of Central and South America between the tropics. South of the Tropic of Cancer the number of species met with falls off very considerably, so that in the Argentine Republic only fourteen members of the family have as yet been recognized. This number will, however, be probably increased when the less-known wooded districts of northern and eastern Argentina have been more fully explored.

The fourteen species of Argentine Tanagers belong to ten different genera, mostly of wide distribution. But to this rule one genus (*Stephanophorus*) is an exception, being only found in Southern Brazil, Paraguay, Uruguay, and Northern La Plata.

---

33. EUPHONIA NIGRICOLLIS (Vieill.).

(BLACK-NECKED TANAGER.)

Euphonia nigricollis, *Scl. Cat. B.* xi. p. 61. **Lindo azul y oro cabeza celeste,** *Azara, Apunt.* i. p. 390. **Euphonia aureata,** *d'Orb. Voy., Ois.* p. 267 (Corrientes).

Description.—Above glossy purplish black; rump yellow; cap and nape blue; front black; below orange-yellow, throat and sides of the head black; axillaries pale yellow; under wing-coverts whitish, inner margins of wing-feathers pale cinereous; bill black; feet pale brown: whole length 4·5 inches, wing 2·7, tail 1·5. *Female* above olive-green, rather lighter on the rump; cap blue; front chestnut, with a narrow dark margin between it and the blue cap; below yellowish olive-green, brighter on the belly.

Hab. South America from Colombia to Northern Argentina.

This Tanager was obtained by d'Orbigny in Southern Corrientes.

---

34. EUPHONIA CHLOROTICA (Linn.).

(PURPLE-AND-YELLOW TANAGER.)

Euphonia chlorotica, *Scl. et Salv. Nomencl.* p. 17; *White, P. Z. S.* 1882, p. 596 (Misiones, Catamarca); *Scl. Cat. B.* xi. p. 64.

Description.—Above dark purple-black; front half of the cap yellow; below yellow, throat purple-black; tail beneath black, with a large white patch on the inner webs of the two exterior tail-feathers; under surface of wings black, with a large white patch on the inner webs of the remiges; bill and feet black: whole length 3·3 inches, wing 2·1, tail 1·2. *Female* above greyish olive-green, with a yellowish tinge on the front and rump; below rather more yellowish, with the centre of the breast and belly pale ashy, flanks and crissum pale yellow; under wing-coverts and inner margins of wing-feathers whitish.

Hab. Guiana, Brazil, Paraguay, and Northern Argentina.

White obtained examples of this species at Concepcion and Catamarca. At the former place it was noticed feeding on the fruit of a giant cactus.

---

35. PIPRIDEA MELANONOTA (Vieill.).

(DARK-BACKED TANAGER.)

Pipridea melanonota, *Scl. Cat. B.* xi. p. 92. **Pica de punza azul y canela,** *Azara, Apunt.* i. p. 413.

Description.—Above violaceous blue; whole interscapular region darker, blackish; wings and tail black, edged with blue; narrow front, lores, and sides of the head deep velvety black, well defined; body beneath and under wing-coverts clear ochraceous; under surfaces of wings and tail blackish; bill black;

feet brown: whole length 5·8 inches, wing 3·1, tail 2·2. *Female* above dark brown, tinged with blue on the head and rump; below like the male.

Hab. South America from Venezuela to Paraguay and Northern Argentina.

This Tanager, which was found by Azara in Paraguay, occurs in Tucuman. An adult male obtained by Herr Schulz in this province is in the collection of Hans, Graf von Berlepsch.

---

36. STEPHANOPHORUS LEUCOCEPHALUS (Vieill.).

(WHITE-CAPPED TANAGER.)

[Plate IV.]

Stephanophorus leucocephalus, *Scl. et Salv. Nomencl.* p. 20; *Durnford, Ibis,* 1877, p. 170 (Buenos Ayres); *White, P. Z. S.* 1882, p. 597 (Misiones); *Barrows, Bull. Nutt. Orn. Cl.* viii. p. 90 (Concepcion); *Scl. Cat. B.* xi. p. 143.
Stephanophorus cœruleus, *Burm. La-Plata Reise,* ii. p. 480 (Paraná).

Description.—Uniform deep blue; cap silky white, with a small crimson crest; bill brownish black, feet brown: whole length 7 inches, wing 4, tail 3·3. *Female* similar, but not quite so bright in colour.

Hab. Southern Brazil, Paraguay, and Northern Argentina.

Azara gave the generic name *Lindo* (beautiful) to the Tanagers, and this species he named the "Blue White-headed Beautiful," the entire plumage being of a very lovely deep corn-flower blue, except a cap of silvery-white feathers on the head, with a crimson spot on the forehead, looking like a drop of blood.

It is a summer bird in Buenos Ayres, where it makes its appearance in spring in the woods bordering on the Plata river, and is usually seen singly or in pairs. The nest is built in a tree ten or twelve feet from the ground, and is somewhat shallow and lined with soft dry grass. The female lays four eggs, white and spotted with deep red. During incubation the male sits concealed in the thick foliage close by, amusing itself by the hour with singing, its performance consisting of chattering disconnected notes uttered in so low a tone as to make one fancy that the bird is merely trying to recall some melody it has forgotten, or endeavouring to construct a new one by jerking out a variety of sounds at random. The bird never gets beyond this unsatisfactory stage, however, and must be admired for its exquisite beauty alone.

ARG. ORN. PL. IV

STEPHANOPHORUS LEUCOCEPHALUS.

37. TANAGRA SAYACA, Linn.
(BLUE TANAGER.)

Tanagra cyanoptera, *Durnford, Ibis,* 1878, p. 59 (Buenos Ayres); *Barrows, Bull. Nutt. Orn. Cl.* viii. p. 91 (Concepcion); *Scl. Cat. B.* xi. p. 157 (part.). **Tanagra sayaca,** *Burm. La-Plata Reise,* ii. p. 479 (Paraná); *Berl. Zeitschr. ges. Orn.* 1885, p. 119; *Scl. Cat. B.* xi. p. 158.

Description.—Bluish grey, paler below; wings and tail blackish, edged with greenish blue; lesser wing-coverts dull greenish blue, like the edgings of the wings; bill plumbeous, feet brown: whole length 6·0 inches, wing 3·6, tail 2·7. *Female* similar.

Hab. S.E. Brazil and Argentina.

In the 'Catalogue of Birds' I referred the Argentine Blue Tanager to *T. cyanoptera,* though expressing a doubt upon the subject. I am now disposed to adopt Graf. v. Berlepsch's view that the Argentine species rather belongs to *T. sayaca.*—P. L. S.

This species migrates as far south as Buenos Ayres, and appears in spring, in small flocks or parties of three or four birds, in the woods on the shores of the Plata. The male utters a series of peculiar squealing notes by way of song.

---

38. TANAGRA BONARIENSIS (Gm.).
(BLUE-AND-YELLOW TANAGER.)

Tanagra bonariensis, *Berl. Zeitschr. ges. Orn.* 1885, p. 119; *Scl. Cat. B.* xi. p. 164. **Tanagra striata,** *Burm. La-Plata Reise,* ii. p. 480 (Buenos Ayres, Mendoza, Cordova); *Scl. et Salv. Nomencl.* p. 21; *Durnford, Ibis,* 1877, p. 170 (Buenos Ayres); *White, P. Z. S.* 1882, p. 596 (Catamarca); *Barrows, Bull. Nutt. Orn. Cl.* viii. p. 91 (Concepcion).

Description.—Lores, region round the eyes, and back black; rump orange; head all round and edging of wings and tail blue; abdomen yellow, passing into orange on the breast; bill horn-colour; feet brown: whole length 7 inches, wing 3·6, tail 2·8. *Female* greyish brown; beneath paler, tinged with yellowish olive on the rump and throat.

Hab. South Brazil, Uruguay, Paraguay, Argentine Republic, and Bolivia.

The plumage of the male of this fine Tanager is rich blue above and yellow beneath; the upper plumage of the female is a uniform dull olive-green, the breast and belly buff-colour.

It is a migratory species, which appears in Buenos Ayres in small flocks in summer. Both sexes have a long, sharp, reedy call-note; the male also possesses a song composed of notes with a peculiar bleating sound.

---

39. PYRANGA AZARÆ, d'Orb.

(AZARA'S TANAGER.)

Pyranga azaræ, *Durnford, Ibis,* 1880, p. 353 (Tucuman); *White, P. Z. S.* 1883, p. 37 (Cordova); *Scl. Cat. B.* xi. p. 186. **Pyranga coccinea,** *Burm. La-Plata Reise,* ii. p. 479 (Paraná, Mendoza). **Pyranga saira,** *Barrows, Bull. Nutt. Orn. Cl.* viii. p. 91 (Concepcion).

Description.—Above dull rosy red; interscapulium and wing-edgings with cinereous tinge; below much brighter, nearly uniform rosy red, duller on the sides; bill plumbeous; feet brown: total length 7·2 inches, wing 3·8, tail 3·1. *Female* greyish olive; beneath yellow, passing into cinereous on the flanks and belly.

Hab. Argentine Republic, Paraguay, Uruguay, and Bolivia.

This scarlet Tanager appears occasionally in the northern and eastern provinces of Argentina.

40. TRICHOTHRAUPIS QUADRICOLOR (Vieill.).

(FOUR-COLOURED TANAGER.)

Trichothraupis quadricolor, *Scl. et Salv. Nomencl.* p. 23; *White, P. Z. S.* 1882, p. 597 (Misiones); *Scl. Cat. B.* xi. p. 220.

Description.—Above greyish olive; a large, recumbent, vertical crest bright yellow; sides of the head, wings, and tail black; below pale fulvous; under wing-coverts and a cross-bar near the base of the wing-feathers white; bill whitish, plumbeous at base; feet pale brown: total length 6·0 inches, wing 3·3, tail 2·9. *Female* similar, but without the vertical crest.

Hab. Southern Brazil, Paraguay, and N.E. Argentina.

This species was met with by White in the dense forests of Misiones.

41. THLYPOPSIS RUFICEPS (d'Orb. et Lafr.).

(RED-CAPPED TANAGER.)

Thlypopsis ruficeps, *Scl. Cat. B.* xi. p. 231.

Description.—Above cinereous; cap bright chestnut-red; beneath yellow, flanks tinged with cinereous; under wing-coverts white; bill plumbeous, feet pale brown: whole length 5 inches, wing 2·5, tail 2·0.

Hab. Bolivia and Tucuman.

Herr Schulz obtained specimens of this Bolivian species in Tucuman.

42. BUARREMON CITRINELLUS, Cab.

(YELLOW-STRIPED TANAGER.)

Buarremon citrinellus, *Cab. Journ. f. Orn.* 1883, p. 109; *Scl. Cat. B.* xi. p. 270.

Description.—Above olive-green, darker on the head; wings and tail blackish brown, edged with olive; a broad superciliary stripe commencing on the front on each side, and another commencing at the gape, bright yellow, leaving in the middle a broad patch of dark olive; beneath yellow, breast and flanks olivaceous; throat yellow, bordered on each side by a dark olive mystacal stripe; bill black; feet brown: whole length 6·5 inches, wing 2·8, tail 2.

Hab. Tucuman.

This is a rather aberrant species of *Buarremon,* as yet only known from Tucuman, where it was discovered by Schulz.

---

43. ARREMON ORBIGNII, Sclater.

(D'ORBIGNY'S TANAGER.)

Arremon orbignii, *Scl. et Salv. Nomencl.* p. 25; *White, P. Z. S.* 1882, p. 597 (Catamarca); *Scl. Cat. B.* xi. p. 274.

Description.—Above olive-green; wings and tail blackish, edged with olive; head black, superciliaries white, vertical stripe and nape cinereous; beneath white, with a narrow but distinct pectoral band; sides cinereous; bend of wing yellow; bill yellow, with the upper half of the upper mandible black; feet brown: total length 5·7 inches, wing 2·8, tail 2·6.

Hab. Bolivia and Northern Argentina.

Specimens of this Tanager were obtained by White on the Sierras of Totoral, near Catamarca, in July 1880. The iris is marked as "blue."

---

44. SALTATOR SIMILIS, d'Orb. et Lafr.

(ALLIED SALTATOR.)

Saltator similis, *d'Orb. Voy., Ois.* p. 290, t. xxviii. fig. 2 (Corrientes); *Scl. Cat. B.* xi. p. 287. **Saltator superciliaris,** *White, P. Z. S.* 1882, p. 597 (Corrientes)?

Description.—Above cinereous; interscapulium and greater wing-coverts and secondaries suffused with olive-green; superciliaries white; below pale fulvous; throat white, bordered on each side by a black line; flanks passing into cinereous; under wing-coverts pale fulvous; bill plumbeous; feet brown: whole length 8·5 inches, wing 4·0, tail 4·0. *Female* similar.

Hab. S.E. Brazil and Eastern Argentina.

This Saltator was first described from specimens obtained at Rincon de Luna in Corrientes by d'Orbigny, but no recent collectors appear to have met with it in the Argentine Republic.

45. SALTATOR CÆRULESCENS, Vieill.

(GREYISH SALTATOR.)

Saltator cærulescens, *Burm. La-Plata Reise*, ii. p. 480 (Paraná); *Scl. et Salv. Nomencl.* p. 26; *Durnford, Ibis*, 1878, p. 59 (Buenos Ayres); *Salvin, Ibis*, 1880, p. 353 (Salta); *White, P. Z. S.* 1882, p. 597 (Salta); *Scl. Cat. B.* xi. p. 290.

Description.—Above cinereous, wings and tail rather darker; short superciliaries white; beneath pale cinereous; throat white, with a blackish rictal stripe on each side; fore neck slightly tinged with fulvous; middle of belly whitish; lower belly and crissum strongly suffused with fulvous; under wing-coverts pale fulvous; bill black; feet brown: total length 8·5 inches, wing 4·3, tail 4·1. *Female* similar.

Hab. Paraguay, Uruguay, and Argentina.

Mr. Durnford found this Saltator "pretty common" near Baradero, on low scrubby ground near water, and afterwards obtained it near Salta. White records it as "very abundant" near Oran.

---

46. SALTATOR AURANTIIROSTRIS, Vieill.

(YELLOW-BILLED SALTATOR.)

Saltator aurantiirostris, *d'Orb. Voy. Ois.* p. 288 (Corrientes); *Burm. La-Plata Reise*, ii. p. 481 (Mendoza, Cordova, Tucuman, Paraná); *Scl. et Salv. Nomencl.* p. 26; *White, P. Z. S.* 1882, p. 598 (Catamarca); *Barrows, Bull. Nutt. Orn. Cl.* viii. p. 91 (Concepcion); *Scl. Cat. B.* xi. p. 292.

Description.—Above cinereous, with a slight olivaceous suffusion; head rather darker; sides of head and throat black; superciliary stripe, commencing above the eye, white; beneath pale cinereous mixed with fulvous, a well-marked guttural collar joining the sides of the throat black; ends of the outer rectrices more or less varied with white; bill clear orange; feet greyish brown: whole length 7·5 inches, wing 3·6, tail 3·5. *Female* similar, less brightly coloured, and with the guttural collar almost or altogether absent.

Hab. Paraguay, Uruguay, and Northern Argentina.

In Corrientes d'Orbigny found this Saltator breeding in the month of November. It frequents the shrubs and bushes in the neighbourhood of the houses, and makes an open nest of roots, not of very neat construction. The eggs are two or three, greenish blue, slightly spotted at the larger end with blackish and reddish zigzag markings. The egg is figured in d'Orbigny's 'Voyage' (pl. xxviii. fig. 3).

White tells us that this species is not uncommon in Catamarca, and Barrows met with it at Concepcion in Entrerios.

Fam. X. FRINGILLIDÆ, or FINCHES.

The extensive family of Finches, which has numerous representatives in every part of the world, is well represented in the Argentine Republic. Within the limits assigned to the present work forty-six species of this group are already known to occur, and it is probable that this number will be somewhat increased when the Argentine avifauna is thoroughly worked out.

Most of the genera to which the Argentine Finches belong are forms peculiar to the Neotropical Region, or at all events to the New World. The genus *Chrysomitris*, which contains our Siskin and other Old-World species, is the only one which has a wider distribution.

One of the most characteristic forms of Argentine Fringillidæ is *Poospiza*, of which seven species are met with within the limits of the Republic, while *Lophospingus*, *Donacospiza*, *Gubernatrix*, and *Saltatricula* are Fringilline types peculiar to Argentina.

---

47. PHEUCTICUS AUREIVENTRIS (d'Orb. et Lafr.).
(BLACK-AND-YELLOW THICK-BILL.)

Pheucticus aureiventris, *Scl. et Salv. Nomencl.* p. 27; *White, P. Z. S.* 1882, p. 598 (Salta); *Sharpe, Cat. B.* xii. p. 55.

Description.—Above, also head, throat, and chest, black; lesser wing-coverts yellow; two spots on the greater coverts and the base of the primaries white; tail black, the three outer feathers on each side tipped with white; belly and under wing-coverts bright yellow; sides sparsely spotted with black; bill and feet black: total length 8·7 inches, wing 4·5, tail 3·5. *Female* similar.

Hab. Bolivia and Northern Argentina.

Examples of this fine species, originally discovered by d'Orbigny in Bolivia, were obtained by White in the forests of Salta.

---

48. GUIRACA CYANEA (Linn.).
(INDIGO FINCH.)

Guiraca cyanea, *Scl. et Salv. Nomencl.* p. 27; *White, P. Z. S.* 1882, p. 598 (Catamarca). **Guiraca cyanea argentina**, *Sharpe, Cat. B.* xii. p. 73.

Description.—Indigo-blue; lesser wing-coverts azure-blue; wing- and tail-feathers black; forehead azure-blue, and head washed with the same colour; bill and feet black: total length 6·8 inches, wing 3·6, tail 2·9. *Female* brown; beneath brighter and rufescent.

Hab. Brazil, Paraguay, and Northern Argentina.

White found this bird "not uncommon around Andalgala in Catamarca, in hedges and thickets." Mr. Sharpe separates the Argentine bird as a subspecies, from its larger size.

49. GUIRACA GLAUCOCÆRULEA (d'Orb. et Lafr.).

(GLAUCOUS FINCH.)

Guiraca glaucocærulea, *Scl. et Salv. P. Z. S.* 1868, p. 139 (Buenos Ayres), *iid. Nomencl.* p. 27; *Durnford, Ibis,* 1877, p. 170 (Buenos Ayres); *Salvin, Ibis,* 1880, p. 353 (Salta); *Barrows, Bull. Nutt. Orn. Cl.* viii. p. 92 (Concepcion); *Sharpe, Cat. B.* xii. p. 75. **Coccoborus glaucocæruleus,** *Burm. La-Plata Reise,* ii. p. 488 (Paraná).

Description.—Uniform glaucous blue; wings and tail blackish, the feathers edged with light blue: total length 5·7 inches, wing 2·9, tail 2·3.

Hab. Paraguay, Uruguay, and Argentina.

This is a rare Finch, and its massive beak and rich blue plumage give to it a highly interesting appearance; but about its habits I have little to tell, for it is essentially a bird of the wild forest, seldom coming near the abodes of man, and being, moreover, shy in disposition, it is difficult to observe it even in its haunts. It is migratory, and is usually seen singly or in pairs, or in small companies of four or five individuals. The male sings, but his performance is merely a confused medley of chattering notes, uttered in so low a tone that they can scarcely be heard at a distance of twelve yards.

50. ORYZOBORUS MAXIMILIANI, Cab.

(PRINCE MAX.'S FINCH.)

Oryzoborus maximiliani, *Burm. La-Plata Reise,* ii. p. 488 (Tucuman); *Sharpe, Cat. B.* xii. p. 78.

Description.—Black; large alar speculum, axillaries, and under wing-coverts white; bill pale yellow; feet brown: total length 5·8 inches, wing 2·9, tail 2·5. *Female* brown; beneath brownish ochraceous.

Hab. S. Brazil, Paraguay, and Northern Argentina.

This Finch is stated by Burmeister to occur near Tucuman.

51. SPERMOPHILA PALUSTRIS, Barrows.

(MARSH FINCH.)

Spermophila palustris, *Barrows, Bull. Nutt. Orn. Cl.* viii. p. 92 (Concepcion); *Sharpe, Cat. B.* xii. p. 112.

Description.—Above, from bill to rump, clear bluish ash; below, from bill to middle of breast, including lower eyelid, ear-coverts, and sides of neck, pure white; rest of underparts, rump, and upper tail-coverts bright chestnut-red; wings and tail blackish brown, edged with whitish; inner secondaries black, their tips and outer edges broadly white; a white patch across the base of all the primaries except the first two; bill and feet black, iris dark: total length 4·50 inches, wing 2·18, tail 1·70. *Female* above uniform greenish olive, obscurely streaked with dusky; below light yellowish buff; wings and tail nearly as in male, but duller.

This small and beautiful Finch was discovered by Mr. Barrows in February 1880, on the Lower Uruguay. It inhabits the marshes, where the males are frequently seen pursuing each other in the pairing-season, occasionally pausing on the top of some tall grass to pour out their delightful song. In character this resembles that of the Black-headed Goldfinch; but has, Mr. Barrows says, a variety and sweetness far beyond the powers of that bird.

It is just possible that Mr. Barrows's Finch may be Azara's unidentified *Chipiu pardo y canela* (Apuntamientos, vol. i. no. 143), which has the same habits, living in the dense reed-beds of the Paraguayan marshes, and is also a delightful singer,—superior to the Goldfinch or Canary, Azara says.

---

52. SPERMOPHILA MELANOCEPHALA (Vieill.).

(BLACK-HEADED FINCH.)

Spermophila melanocephala, *Sharpe, Cat. B.* xii. p. 118.

Description.—Above black; rump and lesser wing-coverts bright rufous; large loral spot on each side white; wings and tail black; alar speculum white; beneath, throat white, band across the breast black; belly pale rufous; flanks and crissum dark rufous; under wing-coverts white; bill and feet black: whole length 5·0 inches, wing 2·4, tail 2·2.

Hab. Paraguay and Argentina.

A specimen of this species, now in the British Museum, was obtained by Durnford at Punta Lara in October 1875.

---

53. SPERMOPHILA CÆRULESCENS (Vieill.).

(SCREAMING FINCH.)

Spermophila cærulescens, *Scl. Ibis,* 1871, p. 12; *Scl. et Salv. Nomencl.* p. 28; *White, P. Z. S.* 1882, p. 598 (Buenos Ayres); *Barrows, Bull. Nutt. Orn. Cl.* viii. p. 92 (Concepcion); *Sharpe, Cat. B.* xii. p. 126. **Sporophila ornata,** *Burm. La-Plata Reise,* ii. p. 488 (Mendoza, Paraná). **Spermophila ornata,** *Scl. et Salv. P. Z. S.* 1869, p. 632; *Durnford, Ibis,* 1877, p. 170 (Buenos Ayres); *Salv. Ibis,* 1880, p. 353 (Salta).

Description.—Above pale smoky brown; front and lores black; beneath, chin and upper part of throat black, with a distinct white mystacal stripe on each side; fore neck white; broad band across the chest black; abdomen white, slightly varied with grey and black on the flanks; under wing-coverts white; bill pale horn-colour; feet brown: whole length 4·8 inches, wing 2·3, tail 1·9. *Female* pale olive-brown; wings and tail darker; beneath lighter, tinged with ochraceous; middle of the belly almost white.

Hab. Southern Brazil, Paraguay, Argentina, and Bolivia.

This species is a summer visitor in Buenos Ayres, and is one of the last to arrive and first to depart of our migrants. These birds are always most abundant in plantations, preferring peach-trees, but do not associate in flocks: they are exceedingly swift and active, overflowing with life and energy, their impetuous notes and motions giving one the idea that they are always in a state of violent excitement. The male has a loud, startled chirp, also a song composed of eight or ten notes, delivered with such vehemence and rapidity, that they run into each other and sound more like a scream than a song. There is not a more clever architect than this species; and while many *Synallaxes* are laboriously endeavouring to show how stately a mansion of sticks a little bird can erect for itself, the Screaming Finch has successfully solved the problem of how to construct the most perfect nest for lightness, strength, and symmetry with the fewest materials. It is a small, cup-shaped structure, suspended hammock-wise between two slender upright branches, and to which it is securely attached by fine hairs and webs. It is made of thin, pale-coloured, fibrous roots, ingeniously woven together—reddish or light-coloured horse-hair being sometimes substituted; and so little material is used that, standing under the tree, a person can easily count the eggs through the bottom of the nest. Its apparent frailness is, however, its best protection from the prying eyes of birds and mammals that prey on the eggs and young of small birds; for it is difficult to detect this slight structure, through which the sunshine and rain pass so freely. So light is the little basket-nest that it may be placed on the open hand and blown away with the breath like a straw; yet so strong that a man can suspend his weight from it without pulling it to pieces. The eggs are three in number, white and spotted with black, sometimes bluish-brown spots are mingled with the black.

54. PAROARIA CUCULLATA (Lath.).

(CARDINAL FINCH.)

Paroaria cucullata, *Burm. La-Plata Reise*, ii. p. 482 (Paraná, Tucuman); *Scl. et Salv. Nomencl.* p. 30; *Durnford, Ibis*, 1877, p. 171 (Buenos Ayres); *Gibson, Ibis*, 1880, p. 19 (Buenos Ayres); *White, P. Z. S.* 1882, p. 598 (Buenos Ayres); *Barrows, Bull. Nutt. Orn. Cl.* viii. p. 129 (Concepcion).

Description.—Above grey; wing- and tail-feathers blackish grey; head all round, crest, and throat brilliant scarlet, the scarlet extending downwards to the chest; below white, the white colour extending up the sides of the neck; nape spotted with white: total length 8·0 inches, wing 4·0, tail 3·5. *Female* similar.

Hab. Bolivia, Paraguay, and Argentina.

This well-known species is perhaps the finest Finch the Argentines have. The entire upper plumage is clear grey, the under surface pure white; but its chief glory is its crest, which, with the anterior part of the head and the throat, is of the most vivid scarlet. The song has little variety, but is remarkably loud, and has that cheerful ring which most people admire in their caged pets, possibly because it produces the idea in the listener's mind that the songster is glad to be a prisoner. As a cage-bird this Finch enjoys an extraordinary popularity; and a stranger in Buenos Ayres, seeing the numbers that are exposed for sale by the bird-dealers in the markets of that city, might fancy that a Cardinal in a cage is considered a necessary part of the *ménage* of every house in the country. This large supply of caged birds comes from South Brazil, Paraguay, and the north-eastern part of the Argentine country, where the Cardinals are most abundant and unite in large flocks. Probably they are not snared, but taken when young from the nest, as most of the birds exposed for sale are in immature plumage.

The Cardinal in a wild state is found as far south as the province of Buenos Ayres, but it is there a scarce bird. It breeds, Mr. Gibson writes, at the end of October, and makes a shallow nest of twigs, vine-tendrils, and horse-hair. The eggs are four; ground-colour white or tinged with faint brown or greenish, and spotted with brown, more densely at the large end.

55. PAROARIA CAPITATA (d'Orb. et Lafr.).

(LESSER CARDINAL FINCH.)

Paroaria capitata, *Burm. La-Plata Reise,* ii. p. 482; *Scl. et Salv. Nomencl.* p. 30; *White, P. Z. S.* 1882, p. 598.

Description.—Above black; head bright scarlet, the black and red divided by white, crescent-shaped marks on the neck, narrowing to a point on the nape; throat black, the black extending downward to the chest; below white; bill and feet yellowish horn-colour: total length 6·6 inches, wing 3·0, tail 2·6.

Hab. Paraguay and Bolivia.

Dr. Burmeister tells us this species is not uncommon near Paraná, where it occurs in small flocks on the river-bank, and is often seen on the stones at the river's edge.

---

56. CORYPHOSPINGUS CRISTATUS (Gm.).

(RED-CRESTED FINCH.)

Coryphospingus cristatus, *Scl. et Salv. Nomencl.* p. 30; *Salvin, Ibis,* 1880, p. 354 (Salta); *White, P. Z. S.* 1882, p. 598 (Salta).

Description.—Above black, washed with red; rump crimson; wing- and tail-feathers black; forehead black; crest vivid scarlet; whole under surface fine crimson: total length 5·5 inches, wing 2·6, tail 2·3. *Female* above brown, with scarlet tinge on the rump; beneath salmon-colour, whitish on the throat.

Hab. Ecuador, Peru, Bolivia, Brazil, Paraguay, and Northern Argentina.

Both Durnford and White obtained specimens of this bird in the province of Salta; White's examples were met with in open country, dotted with thickets of low brushwood, to which the bird resorts.

---

57. LOPHOSPINGUS PUSILLUS (Burm.).

(DARK-CRESTED FINCH.)

Lophospiza pusilla, *Burm. La-Plata Reise,* ii. p. 483. **Lophospingus pusillus,** *Cab. J. f. O.* 1878, p. 195 (Cordova). **Coryphospingus pusillus,** *Salv. Ibis,* 1880, p. 354, pl. ix. fig. 1 (Tucuman).

Description.—Above grey; sides of head and crest blackish; broad superciliaries white; wings blackish, edged with whitish grey; tail blackish, lateral rectrices with broad white tips; beneath greyish white, clearer on the throat and middle of the belly; bill dark horn-colour; lower mandible whitish; feet pale brown: whole length 5·0 inches, wing 2·5, tail 2·4.

Hab. Northern Argentina.

58. DONACOSPIZA ALBIFRONS (Vieill.).

(LONG-TAILED REED-FINCH.)

Donacospiza albifrons, *Scl. et Salv. P. Z. S.* 1869, p. 632 (Buenos Ayres), *iid. Nomencl.* p. 30; *Durnford, Ibis,* 1877, p. 171 (Buenos Ayres); *White, P. Z. S.* 1883, p. 38 (Cordova); *Döring, Exp. al Rio Negro, Zool.* p. 38 (Pampas). **Poospiza albifrons,** *Burm. La-Plata Reise,* ii. p. 484 (Paraná).

Description.—Above yellowish grey, the back striped with blackish; lesser wing-coverts clear grey; greater coverts and wing-feathers black, edged with brown; head nearly the same as the back, somewhat grey on the cheek, the crown and nape washed with olive-brown; superciliary stripe and under surface buff; bill and feet horn-colour: total length 6·0 inches, wing 2·5, tail 3·0.

Hab. Paraguay, Uruguay, and La Plata.

The slender body, great length of tail, and the hue of the plumage, assimilating to that of sere decaying vegetation, might easily lead one into mistaking this Finch for a *Synallaxis* where these birds are abundant.

I have met with it in the marshy woods and reed-beds along the shores of the Plata, but it is a shy, rare bird in Buenos Ayres. I have followed it about, hoping to hear it utter a song or melodious note, but it had only a little chirp. I would not, however, on this account pronounce it to be the one silent member of a voiceful family, as my acquaintance with it is so very slight.

---

59. POOSPIZA NIGRORUFA (d'Orb. et Lafr.).

(BLACK-AND-CHESTNUT WARBLING FINCH.)

Poospiza nigrorufa, *Burm. La-Plata Reise,* ii. p. 484 (Paraná); *Scl. et Salv. Nomencl.* p. 30; *Durnford, Ibis,* 1877, p. 171 (Buenos Ayres); *Döring, Exp. al Rio Negro, Zool.* p. 38 (Rio Sauce); *Barrows, Bull. Nutt. Orn. Cl.* viii. p. 129 (Concepcion).

Description.—Entire upper parts black, faintly washed with olive; superciliary stripe pale straw-colour; two outer tail-feathers on each side tipped with white; throat and under surface bright chestnut; centre of abdomen white; under tail-coverts pale buff: total length 5·8 inches, wing 2·5, tail 2·2. *Female*: upper parts not so dark as in male; underparts light buff, mottled and striped with blackish.

Hab. South Brazil, Paraguay, and Argentina.

This sweet-voiced little songster appears in Buenos Ayres at the end of September; it is a common bird in grounds abounding in bushes and scattered trees, and, in its bright ruddy breast and dark upper plumage, has some resemblance to the English Robin; only it has a very conspicuous straw-coloured line above the eye. Its voice also, in purity and sweetness of tone, is not unlike that of the Robin; but the song, composed of six unvarying notes, is uttered in a deliberate, business-like manner at regular intervals, and is monotonous. Never more than two birds are seen together; they feed on the ground in humid situations, the male frequently seeking a perch to sing. The nest is made on the ground, or in a close bush near the surface; the eggs have a pale bluish ground-colour, irregularly marked with black and very dark brown spots, and in some instances clouded with faint grey.

---

60. POOSPIZA WHITII, Scl.

(WHITE'S WARBLING FINCH.)

Poospiza whitii, *Scl. P. Z. S.* 1883, p. 43, pl. ix.

Description.—Above clear grey, very faintly washed with olive; tail black, the outer feathers on each side tipped with white, as in *P. erythrophrys*; patch between bill and eye, the anterior half of superciliary stripe, and chin white; hinder part of superciliary stripe and under surface chestnut; middle of belly white; under tail-coverts buff: total length 5·5 inches, wing 2·4, tail 2·6. *Female* similar, but breast pale rufous; abdomen white; sides grey and buff.

Hab. Province of Cordova, Rep. Arg.

This species has only been obtained by the late Mr. E. W. White. He met with specimens of it at Cosquin, Cordova, in June, July, and August, 1882. It has been dedicated to its discoverer, who has so largely augmented our knowledge of the Argentine avifauna, and whose premature death was a veritable loss for science.

---

61. POOSPIZA ERYTHROPHRYS, Scl.

(RED-BROWED WARBLING FINCH.)

Poospiza erythrophrys, *Scl. Ibis,* 1881, p. 599, pl. xvii. fig. 1; *White, P. Z. S.* 1882, p. 599 (Catamarca).

Description.—Above grey, faintly washed with olive; greater wing-coverts tipped with white; wing-feathers blackish, edged with grey; tail blackish grey, the two outer feathers on each side tipped with white, and on the outer feather the white extending down the outer web to the base of the tail; superciliary stripe and whole under surface light chestnut, paler on the abdomen; under tail-coverts buff: total length 5·5 inches, wing 2·5, tail 2·3.

Hab. Northern Argentina.

This pretty species is another of White's discoveries. He first met with it in the Sierra of Totoral, Catamarca, and subsequently about 20 miles north of Buenos Ayres.

---

62. POOSPIZA ASSIMILIS, Cab.

(RED-FLANKED WARBLING FINCH.)

Poospiza assimilis, *Cab. Mus. Hein.* i. p. 137; *White, P. Z. S.* 1882, p. 599 (Misiones). **Poospiza lateralis,** *Barrows, Bull. Nutt. Orn. Cl.* viii. p. 130 (Concepcion).

Description.—Above grey, slight superciliary mark whitish; middle of back and wings washed with rufous; rump bright rufous; tail blackish, two lateral pairs of rectrices broadly tipped with white; beneath grey, white in the middle of the belly; flanks and crissum bright rufous; under wing-coverts greyish white; bill horn-colour, lower mandible yellowish; feet pale brown: whole length 5·0 inches, wing 2·5, tail 2·4.

Hab. Northern Argentina.

White found this species abundant amongst the thick weeds and grass in the outskirts of Concepcion. He took it for *P. thoracica,* from which it differs in its rufous rump. It is more like *P. lateralis.*

---

63. POOSPIZA ORNATA (Landb.).

(PRETTY WARBLING FINCH.)

Phrygilus ornatus, *Landb. Journ. f. Orn.* 1865, p. 405. **Poospiza ornata,** *Scl. et Salv. Nomencl.* p. 30.

Description.—Above grey; back more or less varied with chestnut; long superciliaries ochreous white; wings blackish, coverts edged with white; tail black, lateral rectrices broadly tipped with white; beneath dark chestnut, lighter on the middle of the belly; bill and feet dark brown: whole length 5·2 inches, wing 2·4, tail 2·3. *Female* similar, but paler in colour, especially below.

Hab. Mendoza.

---

64. POOSPIZA TORQUATA (d'Orb. et Lafr.).

(RINGED WARBLING FINCH.)

Poospiza torquata, *Burm. La-Plata Reise,* ii. p. 484 (Mendoza); *Scl. et Salv. Nomencl.* p. 30; *White, P. Z. S.* 1882, p. 599 (Santiago del Estero); *Döring, Exp. al Rio Negro, Zool.* p. 38 (R. Colorado).

Description.—Above grey; wings blackish, the greater wing-coverts tipped with white; tail-feathers black, the two outer on each side tipped with white; from the bill, a broad white stripe extends above the eye to the nape; crown grey; sides of head black; beneath white, the chest crossed with a broad black band; under tail-coverts rufous: total length 5·3 inches, wing 2·5, tail 2·3.

Hab. Bolivia and Argentina.

White met with a male of this species in September 1881, at San Pedro, in the province of Santiago. Burmeister found it near Mendoza in the valleys of the Sierra de Uspallata.

---

65. POOSPIZA MELANOLEUCA (Vieill.).

(WHITE-AND-GREY WARBLING FINCH.)

Poospiza melanoleuca, *Burm. La-Plata Reise,* ii. p. 484 (Entrerios); *Scl. et Salv. Nomencl.* p. 30; *Salv. Ibis,* 1880, p. 354, pl. ix. fig. 2 (Tucuman); *White, P. Z. S.* 1882, p. 599 (Salta); *Barrows, Bull. Nutt. Orn. Cl.* viii. p. 130 (Entrerios).

Description.—Above, head and neck blackish; back grey; wings and tail blackish, three outer rectrices on each side tipped with white; beneath white; sides washed with grey, breast and belly also faintly tinged with grey: whole length 5·1 inches, wing 2·3, tail 2·0. *Female* similar.

Hab. Argentina.

White says that in Salta "this bird frequents the gardens in the vicinity of the houses, where it vigorously pursues the ants amongst the branches of the trees." Mr. Barrows says that near Concepcion it had the general motions and appearance of a Titmouse, thus differing widely from the other members of the genus.

---

66. PHRYGILUS GAYI (Eyd. et Gerv.).

(GAY'S FINCH.)

Phrygilus gayi, *Burm. La-Plata Reise,* ii. p. 487 (Mendoza); *Scl. et Salv. Nomencl.* p. 31; *White, P. Z. S.* 1882, p. 599 (Catamarca).

Description.—Above, head all round, neck, and wing-coverts grey; back and upper tail-coverts olive-green; wings and tail black; beneath, throat grey, breast and belly yellow, washed with olive; crissum and under tail-coverts white; bill and feet black: whole length 6·0 inches, wing 3·5, tail 2·3. *Female*: above obscure olive-green, below yellow.

Hab. Chili and Western Argentina.

Burmeister says that this Chilian species is found in the valleys of the Sierras near Mendoza.

---

67. PHRYGILUS CANICEPS (Burm.).

(GREY-HEADED FINCH.)

Phrygilus caniceps, *Burm. La-Plata Reise*, ii. p. 487; *Durnford, Ibis*, 1878, p. 393 (Chupat); *Barrows, Bull. Nutt. Orn. Cl.* viii. p. 130 (Sierra de la Ventana).

Description.—Above, head, neck, wing- and upper tail-coverts grey; back olivaceous yellow, washed with red; wing and tail-feathers black, edged with grey; beneath, throat grey, breast and belly orange-yellow; crissum and under tail-coverts white; bill horn-colour; feet light brown: whole length 6·8 inches, wing 3·6, tail 2·9.

Hab. Argentina.

Of this species, discovered by Dr. Burmeister near Mendoza, Mr. Barrows writes:—"Only met with on the Sierra de la Ventana, where it was abundant in flocks, some of which numbered as many as a hundred individuals. When found near the base of the sierra they were almost always associated with the common *Zonotrichia*. Although most of them had not finished moulting, they were constantly singing, and seemed perfectly contented with their desolate surroundings."

---

68. PHRYGILUS DORSALIS, Cab.

(RED-BACKED FINCH.)

Phrygilus dorsalis, *Cab. Journ. f. Orn.* 1883, p. 109.

Description.—Ashy grey; back rusty red; wing-coverts blackish; chin, lower belly, and crissum whitish: whole length 6·5 inches, wing 3·7, tail 2·6.

Hab. Tucuman.

Schulz discovered this species on the Cerro Vayo of Tucuman, near the snow-line. It reminds one of the North-American species of *Junco* in its coloration.

---

69. PHRYGILUS UNICOLOR (d'Orb. et Lafr.).

(SLATY FINCH.)

Phrygilus unicolor, *Scl. et Salv. Nomencl.* p. 31; *Cab. J. f. O.* 1878, p. 195 (Cordova). **Phrygilus rusticus,** *Burm. La-Plata Reise*, ii. p. 487 (Mendoza).

Description.—Above nearly uniform slaty grey, below paler, whiter on the middle of the belly; bill dark horn-colour, feet clear brown: whole length 6·0 inches, wing 3·6, tail 2·6. *Female* cinereous, with blackish shaft-spots above and below; paler on the middle of the belly.

Hab. Andes of Colombia, Ecuador, and Peru; Chili, Bolivia, and Northern Argentina.

70. PHRYGILUS FRUTICETI (Kittl.).

(MOURNING FINCH.)

Phrygilus fruticeti, *Burm. La-Plata Reise,* ii. p. 487 (Cordillera of Mendoza); *Scl. et Salv. Nomencl.* p. 31; *Hudson, P. Z. S.* 1872, p. 537 (Rio Negro); *Durnford, Ibis,* 1878, p. 393 (Chupat); *Döring, Exp. al Rio Negro, Zool.* p. 39 (R. Colorado, R. Negro). **Emberiza luctuosa,** *Eyd. et Gerv. Mag. de Zool.*, 1836, *Ois.* pl. 72.

Description.—Grey, with minute black markings on the head and neck, and broader stripes on the back; greater coverts and wings black, the feathers edged with grey, and a band of white spots across the coverts; tail black; beneath, throat, and upper part of breast black, many feathers tipped with grey, giving the chest a mottled appearance; lower part of breast and belly grey, mottled below the chest with a few black spots; centre of abdomen and under tail-coverts white; beak yellow; feet flesh-colour: total length 7·3 inches, wing 4·0, tail 3·2. *Female* obscure grey and without the black colour on the throat and chest.

Hab. Peru, Bolivia, Chili, and Argentine Republic.

This Finch is common on the western slopes of the Andes as far north as Peru; it is also found in the Mendoza district and throughout Patagonia. It is very abundant on the Rio Negro, especially in the immediate neighbourhood of the Carmen settlements, for, like the Chingolo and other fringilline species, it is beneficially affected by cultivation. Though not possessing any bright tints, it is a very charming bird, tuneful, elegant in form, graceful and buoyant in its motions. When approached it utters a series of low ticking sounds, and at intervals a peculiar long squealing note. The song of the male is very agreeable, and curiously resembles that of the Cachila Pipit (*Anthus correndera*). It usually sits on a twig near the ground, and at intervals soars up to a height of ten or twenty yards, and utters its song while gliding slowly downwards with depressed wings and outspread tail. It sings throughout the year; in bright weather its notes are heard all day long, but on cold, cloudy, or wet days only after sunset. In the warm season they live in pairs, and in the autumn unite in flocks of as many as two or three hundred individuals, and have a strong undulating flight.

---

71. PHRYGILUS CARBONARIUS (d'Orb. et Lafr.).

(BLACKISH FINCH.)

Phrygilus carbonarius, *Burm. La-Plata Reise,* ii. p. 487; *Döring, Expl. al Rio Negro, Zool.* p. 38.

Description.—Above grey, with brownish-black stripes; wings dusky brown;

tail-feathers black; below black, with minute grey marks on the throat and bosom; sides dull grey: total length 5·5 inches, wing 2·9, tail 2·2.

Hab. Bolivia, Argentina, and Patagonia.

This species is said to be about one third less than *Phrygilus fruticeti* in size, but to resemble it closely in colour. Dr. Döring found it abundant in Patagonia between the rivers Colorado and Negro, where it was seen during the cold season in flocks, associating with *Zonotrichia canicapilla.* Burmeister met with it in the Sierra de Uspallata, near Mendoza.

---

72. GUBERNATRIX CRISTATELLA (Vieill.).

(YELLOW CARDINAL.)

Gubernatrix cristatella, *Burm. La-Plata Reise,* ii. p. 482 (Paraná); *Scl. et Salv. Nomencl.* p. 31; *White, P. Z. S.* 1883, p. 38 (Cordova); *Barrows, Bull. Nutt. Orn. Cl.* viii. p. 130 (Concepcion, Entrerios).

Description.—Above light olive-green, the back marked with a few black stripes; four middle tail-feathers black, edged with olive, all the rest of the tail-feathers bright yellow, tipped with dull olive; broad superciliaries and lower part of the head bright yellow; crown, crest, and throat velvet-black, the black extending to the chest; beneath yellow, washed with olive-green on the breast and sides; bill and feet black: total length 8·0 inches, wing 3·8, tail 3·7. *Female* less brightly coloured; white on the head where the male is yellow; breast grey.

Hab. Paraguay and Argentine Republic.

The Yellow Cardinal is a graceful sprightly bird, with a strong melodious voice, and is one of our favourite cage-birds. It visits Buenos Ayres in small flocks in spring, but is a rare bird with us. There is little variety in its song, which is composed of four or five mellow notes of great power, and in tone somewhat like the whistle of the Blackbird of Europe.

---

73. DIUCA GRISEA (Less.).

(DIUCA FINCH.)

Diuca vera, *Burm. La-Plata Reise,* ii. p. 483 (Mendoza). **Diuca grisea,** *Scl. et Salv. Nomencl.* p. 31.

Description.—Above grey, sides of head darker; wings and tail blackish, edged with grey; lateral rectrices tipped with white; below grey, whole chin and throat and middle of the belly white; lower flanks and crissum stained with rufous; bill plumbeous; feet dark hazel: whole length 6·3 inches, wing 3·3, tail 2·6. *Female* similar, but tinged with brownish, and colours more obscure.

Hab. Chili and Western Argentina.

This well-known Chilian species is said by Burmeister to be not unfrequent near Mendoza and along the range of the Cordilleras.

---

74. DIUCA MINOR, Bp.

(LESSER DIUCA FINCH.)

Diuca minor, *Burm. La-Plata Reise,* ii. p. 483 (San Louis, Cordova); *Scl. et Salv. Nomencl.* p. 31; *Hudson, P. Z. S.* 1872, p. 537 (Rio Negro); *Durnford, Ibis,* 1878, p. 393 (Tombo Point, Pat.); *Döring, Exp. al Rio Negro, Zool.* p. 39 (R. Colorado, R. Nigro); *Barrows, Bull. Nutt. Orn. Cl.* viii. p. 131 (Concepcion, Entrerios).

Description.—Grey; head, neck, and back faintly washed with brown; wings blackish, the feathers edged with brownish grey; tail-feathers black, tipped with white on the inner webs; beneath, from the bill to the chest, white; upper part of breast and sides grey; rest of the under surface white; a bright chestnut spot on the flanks: whole length 6·5 inches, wing 3·2, tail 2·6. *Female* similar, but less bright than male.

Hab. Argentina and Patagonia.

This pretty little grey-and-white Finch is common on the Chilian side of the Andes and throughout Patagonia, and also occurs in the Mendoza district. It is a tuneful bird, lively, social, and frolicsome in disposition; in autumn and winter uniting in flocks of from fifty to three or four hundred individuals; swift of flight, and when on the wing fond of pursuing its fellows and engaging in mock battles. The song of the male is very pleasing, the voice having more depth and mellowness than is usual with the smaller fringilline singers, which, as a rule, have thin, reedy, and tremulous notes. In summer it begins singing very early, even before the faintest indication of coming daylight is visible, and at that dark silent hour the notes may be heard at a great distance and sound wonderfully sweet and impressive. During the cold season, when they live in companies, the singing-time is in the evening, when the birds are gathered in some thick-foliaged tree or bush which they have chosen for a winter roosting-place. This winter-evening song is a hurried twittering, and utterly unlike the serene note of the male bird heard on summer mornings. A little while after sunset the flock bursts into a concert, which lasts several minutes, sinking and growing louder by turns, and during which it is scarcely possible to distinguish the notes of individuals. Then follows an interval of silence, after which the singing is again renewed very suddenly and as suddenly ended. For an hour after sunset, and when all other late singers, like the *Mimus,* have long been silent, this fitful impetuous singing is continued. Close by a house on the Rio Negro, in which I spent several months, there were three very large chañar bushes, where a multitude of Diuca Finches used to roost, and they never missed singing in the evening, however cold or rainy the weather happened to be. So fond were they of this charming habit, that when I approached the bushes or stood directly under them, the alarm caused by my presence would interrupt the performance only for a few moments, and presently they would burst into song again, the birds all the time swiftly pursuing each other amongst the foliage, often within a foot of my head.

The eggs, Darwin says (Zool. Voy. 'Beagle,' iii. p. 93), are pointed, oval, pale dirty green, thickly blotched with pale dull brown, becoming confluent and entirely coloured at the broad end.

---

75. CATAMENIA ANALIS (d'Orb. et Lafr.).

(RED-STAINED FINCH.)

Catamenia analis, *Burm. La-Plata Reise*, ii. p. 488 (Mendoza); *Scl. et Salv. Nomencl.* p. 31; *White, P. Z. S.* 1882, p. 599 (Catamarca). **Spermophila analis,** *Sharpe, Cat. B.* xii. p. 106.

Description.—Above clear grey; wing-feathers black, edged with grey; tail black, a large white blotch on the central part of each feather, the two middle feathers excepted; beneath grey, palest on the belly; under tail-coverts rufous: whole length 5·0 inches, wing 2·8, tail 2·2. *Female*, above obscure brownish buff, striped with blackish; beneath dirty white.

Hab. Peru, Bolivia, and Argentina.

Burmeister met with this Finch on the sierras near Mendoza, and White obtained a single specimen in Catamarca.

---

76. CATAMENIA INORNATA (Lafr.).

(PLAIN-COLOURED FINCH.)

Sporophila rufirostris, *Landb. J. f. O.* 1865, p. 404 (Mendoza). **Catamenia inornata,** *Scl. et Salv. Nomencl.* p. 31. **Spermophila inornata,** *Sharpe, Cat. B.* xii. p. 104.

Description.—Above dull grey, clearer on the rump; wings and tail blackish, wing-feathers edged with grey; beneath grey, under tail-coverts bright chestnut; bill red; feet brown: whole length 5·0 inches, wing 2·5, tail 2·2.

Hab. Bolivia and N. Argentina.

Examples of this species were obtained by Weisshaupt near Mendoza in 1871.

---

77. ZONOTRICHIA PILEATA (Bodd.).

(CHINGOLO SONG-SPARROW.)

Zonotrichia pileata, *Scl. et Salv. P. Z. S.* 1868, p. 139, *iid. Nomencl.* p. 31; *Salvin, Ibis,* 1880, p. 355 (Salta); *Gibson, Ibis,* 1880, p. 28 (Buenos Ayres); *White, P. Z. S.* 1882, p. 600 (Buenos Ayres); *Barrows, Bull. Nutt. Orn. Cl.* viii. p. 131 (Concepcion). **Zonotrichia matutina,** *Burm. La-Plata Reise,* ii. p. 486.

Description.—Above dusky grey, striped with blackish brown; the top of the head from the bill to the nape grey; a whitish stripe from the eye to the nape; between the stripe and the grey on the crown black; a narrow chestnut ring round the neck, widening to a large patch on the sides of the chest, the patch bordered with black on its lower part; beneath, throat white; breast and belly ashy white; bill and feet light horn-colour: whole length 5·7 inches, wing 2·8, tail 2·2. *Female* similar, but duller in colour and a trifle smaller.

Hab. Central and South America.

The common, familiar, favourite Sparrow over a large portion of the South-American continent is the "*Chingolo.*" Darwin says that "it prefers inhabited places, but has not attained the air of domestication of the English Sparrow, which bird in habits and general appearance it resembles." As it breeds in the fields on the ground, it can never be equally familiar with man, but in appearance it is like a refined copy of the burly English Sparrow—more delicately tinted, the throat being chestnut instead of black; the head smaller and better proportioned, and with the added distinction of a crest, which it lowers and elevates at all angles to express the various feelings affecting its busy little mind.

On the treeless desert pampas the Chingolo is rarely seen, but wherever man builds a house and plants a tree there it comes to keep him company, while in cultivated and thickly settled districts it is excessively abundant, and about Buenos Ayres it literally swarms in the fields and plantations. They are not, strictly speaking, gregarious, but where food attracts them, or the shelter of a hedge on a cold windy day, thousands are frequently seen congregated in one place; when disturbed, however, these accidental flocks immediately break up, the birds scattering abroad in different directions.

The Chingolo is a very constant singer, his song beginning with the dawn of day in spring, and continuing until evening; it is very short, being composed of a chipping prelude and four long notes, three uttered in a clear thin voice, the last a trill. This song is repeated at brief intervals, as the bird sits motionless, perched on the disc of a thistle-flower, the summit of a stalk, or other elevation; and where the Chingolos are very abundant, the whole air, on a bright spring morning, is alive with their delicate melody; only one must pause and listen before he is aware of it, otherwise it will escape him, owing to its thin ethereal character, the multitudinous notes not mingling but floating away, as it were, detached and scattered, mere gossamer webs of sound that very faintly impress the sense. They also sing frequently at night, and in that dark silent time their little melody sounds strangely sweet and expressive. The song varies greatly in different districts; thus, in Bahia Blanca it is without the long trill at the end, and in other localities I have found it vary in other ways.

The Chingolos pair about the end of September, and at that time their battles are frequent, as they are very pugnacious. The nest is made under a thistle or tuft of grass, in a depression in the soil, so that the top of the nest is on a level with the surface of the ground. The nest is mostly made and lined with horse-hair, the eggs four or five, pale blue, and thickly spotted with dull brown. Sometimes, though very rarely, a nest is found in a bush or on a stump several feet above the ground. Two broods are reared in the season, the first in October, the second in February or March. I have known these birds to breed in April and May, and these very late nests escape the infliction of parasitical eggs. When the nest is approached or taken, the Chingolos utter no sound, but sit in dumb anxiety, with tail expanded and drooping wings.

---

78. ZONOTRICHIA CANICAPILLA, Gould.

(PATAGONIAN SONG-SPARROW.)

Zonotrichia canicapilla, *Durnford, Ibis,* 1877, p. 33 (Chupat), et 1878, p. 393 (Centr. Patag.); *Sclater, Ibis,* 1877, p. 46, pl. 1. fig. 1; *Döring, Exp. al Rio Negro, Zool.* p. 39 (R. Colorado, R. Negro).

Description.—Head grey, with narrow white superciliaries; in other respects like *Zonotrichia pileata*: total length 6·3 inches, wing 3·2, tail 2·6.

Hab. Patagonia.

Durnford found this species common and abundant on the Chupat River and in the interior of Patagonia. It has a pretty song, and sings in the evening and during the night when the moon is shining. It nests among coarse grass and herbage, making an unpretending structure of the former material, which is lined with fibres. It lays four eggs, pale green, thickly striated with light reddish-brown spots running into each other, and most numerous at the large end.

---

79. ZONOTRICHIA STRIGICEPS, Gould.

(STRIPE-HEADED SONG-SPARROW.)

Zonotrichia strigiceps, *Burm. La-Plata Reise,* ii. p. 486 (Paraná, Santa Fé); *Scl. et Salv. Nomencl.* p. 31; *Scl. Ibis,* 1877, p. 47, pl. 1. fig. 2.

Description.—Above light brownish grey, striped with black; centre of crown ash-grey, under the grey a broad rufous stripe, beneath which is a narrow grey superciliary stripe; behind the eye a rufous mark; beneath, throat white, breast pale grey; sides and belly yellowish grey; middle of belly white: whole length 6·2 inches, wing 2·6, tail 2·6.

Hab. Argentina and Patagonia.

---

80. ZONOTRICHIA HYPOCHONDRIA (d'Orb. et Lafr.).

(RED-FLANKED SONG-SPARROW.)

Emberiza hypochondria, *d'Orb. Voy., Ois.* p. 361, t. 45. fig. 1. **Zonotrichia hypochondria,** *Burm. La-Plata Reise,* ii. p. 486 (Mendoza).

Description.—Brownish grey, head darker; superciliaries white; wings brownish black, edged with greyish rufous; tail brownish black, four external pairs of rectrices with a long white mark on the inner web, the outer pair with the outer web also margined with white; beneath, throat and neck white; sides of head, mystacal line, neck and breast-band plumbeous; belly dirty white; flanks chestnut: whole length 6·0 inches, wing 2·8, tail 2·0.

Hab. Bolivia and Western Argentina.

Prof. Burmeister, who met with this species near Mendoza, says it is a true *Zonotrichia,* and not a *Poospiza,* as sometimes considered.

---

81. COTURNICULUS PERUANUS (Bp.).

(YELLOW-SHOULDERED SONG-SPARROW.)

Coturniculus manimbe, *Burm. La-Plata Reise,* ii. p. 486 (Paraná); *White, P. Z. S.* 1882, p. 600 (Corrientes); *Döring, Exp. al Rio Negro, Zool.* p. 40 (R. Colorado); *Barrows, Bull. Nutt. Orn. Cl.* viii. p. 131 (Concepcion). **Coturniculus peruanus,** *Scl. et Salv. Nomencl.* p. 32.

Description.—Above grey, mottled with rufous-brown; wing-feathers black, edged with rufous; tail-feathers black, edged with dull grey; a patch between the bill and eye and the shoulders bright yellow; beneath, throat whitish; breast and belly and sides dull grey, white on the middle of the belly; bill and feet horn-colour: whole length 4·9 inches, wing 2·4, tail 1·9. *Female* similar, but less bright, the yellow spot on the head scarcely perceptible.

Hab. Peru, Bolivia, and Argentina.

SALTATRICULA MULTICOLOR.

The prevailing colour of this little field-sparrow is grey, marked and mixed with fuscous and brown; the shoulder and space between the beak and eye are yellow. It is a common species in the northern portion of the Argentine country, and appears now to be gradually extending its range southwards. Many years ago I first noticed it on the pampas north of Buenos Ayres; afterwards I found it in the immediate neighbourhood of that city; then it began to spread over the plains to the south, appearing every spring in greater numbers, but it is still far from common. It has, I fancy, a limited migration, as I could never find one in winter. It is solitary, and frequents open plains and fields; lives on the ground, and never alights on a tree. The male has a favourite perch, a tall weed or post, where he spends a great deal of his time, repeating his song at intervals of half a minute; it is short and pleasing, and has a slight resemblance to the song of the Yellow-Hammer, but is more delicate and melodious. When approached, the bird flies down and conceals itself in the grass.

---

82. SALTATRICULA MULTICOLOR, Burm.

(MANY-COLOURED GROUND-FINCH.

[PLATE V.]

Saltatricula multicolor, *Burm. La-Plata Reise,* ii. p. 481 (Paraná); *Scl. et Salv. Nomencl.* p. 32; *Salvin, Ibis,* 1880, p. 355 (Salta); *White, P. Z. S.* 1882, p. 600 (Catamarca).

Description.—Above grey, faintly tinged with olivaceous on the head, neck, and back; a short white stripe behind the eye; beneath the stripe and lower part of neck velvet-black; wings blackish; tail-feathers black, tipped with white; beneath, throat white, breast grey; sides and belly pale chestnut; middle of belly and under tail-coverts white; bill and feet horn-colour: whole length 6·6 inches, wing 2·9, tail 3·0. *Female* similar, but not so bright.

Hab. North and Western Argentina.

To Dr. Burmeister we owe the discovery of this brilliant Finch, as of so many other Argentine species. He met with it at Paraná, in the bushy lands east of the city, and obtained five specimens. White found it near Andagala in Catamarca, on trees, feeding on the insects in the flowers and on seeds; and Durnford collected specimens near Salta.

Examples of this species were also obtained by Weisshaupt near Mendoza (*cf.* Salvin, Ibis, 1880, p. 355).

---

83. EMBERNAGRA PLATENSIS (Gm.).

(RED-BILLED GROUND-FINCH.)

Embernagra platensis, *Burm. La-Plata Reise,* ii. p. 485 (Paraná); *Scl. et Salv. P. Z. S.* 1868, p. 140 (Buenos Ayres), 1872, p. 548 (Rio Negro), *iid. Nomencl.* p. 32; *Durnford, Ibis,* 1877, p. 172 (Buenos Ayres); *White, P. Z. S.* 1882, p. 600 (Buenos Ayres, Corrientes); *Döring, Exp. al Rio Negro, Zool.* p. 40 (R. Colorado, R. Negro); *Barrows, Bull. Nutt. Orn. Cl.* viii. p. 132 (Concepcion, Entrerios).

Description.—Above dull olive-green, striped with blackish; wings silky olive-green, the inner webs of the feathers black; edge of wings yellow; tail-feathers dull olive-green; beneath, throat and breast grey; belly buff; beak bright red; feet pinkish horn-colour: total length 8·8 inches, wing 3·7, tail 3·8. *Female* similar.

Hab. South Brazil, Paraguay, and Argentina.

In this Finch the plumage is alike in both sexes. Above it is dusky olive-green, beneath grey; the beak is of a fine bright red. In Argentina this bird is most common in the littoral forests along the Plata, but ranges as far south as the Rio Negro in Patagonia. It does not migrate, nor associates in flocks; but the sexes are faithful, and the male and female are invariably together, and appear to be very fond of each other's society. They have a loud, sharp alarm chirp or cry, which bursts from the bird with the startling suddenness of a sneeze from a human being; also a confused unmelodious song, which always reminds me, in its hurry, vehemence, and peculiar sound, of the gobbling of a turkey-cock. They are not shy, but when approached sit jerking their tails about, and uttering loud chirps as if greatly excited. The flight is very curious; the bird springs up with great suddenness, and with tail erect, and the long legs dangling down like a Rail's, proceeds by a series of irregular jerks, violently shutting and opening its wings. They breed on the ground under the grass, and conceal their nest so well that I doubt whether the parasitical *Molothrus* ever finds it. I have, at all events, never seen them followed by the young of *Molothrus* demanding food.

As a rule, small seed-eating birds are beneficially affected by the presence of man; thus our common *Zonotrichia* and other sparrows and finches have become excessively numerous in the most thickly-settled districts. With the Red-billed Finch, however, just the contrary has happened; and since I have known this species it has disappeared from many localities where it was once quite common. Azara's name for this species, *Habia de bañado,* signifies that it is a marsh bird; but though now found chiefly in marshy situations, it was once common enough over the entire pampas region, before the great plains were settled on by Europeans. The bird is very badly protected by nature against raptorial species, owing to its very conspicuous red beak, its habit of perching on the summit of tall plants and other elevated positions, its loud impetuous voice, which invites attention, and the weak eccentric flight, which challenges pursuit. It is essential to its safety that it should have, in the open country it frequents, a dense grass cover into which it can plunge on the slightest alarm. Where cattle are introduced, the original pampas-grass, which afforded the suitable conditions, disappears, giving place to the soft, perishable grasses, clovers, and thistles of Europe. Where these changes take place, the bird cannot escape from its enemies and quickly disappears; while many Dendrocolaptine species inhabiting the same situations are saved by their inconspicuous protective colouring, sharp wedge-like bodies, and swift mouse-like motions on the ground. In marshy places on the pampas, abounding with long aquatic grasses and reed-beds, the Red-bill still maintains its existence, but from its old habitat on the open grassy plains, where it was once the dominant Finch, it has utterly vanished.

---

84. EMBERNAGRA OLIVASCENS (d'Orb. et Lafr.).

(OLIVE GROUND-FINCH.)

Embernagra olivascens, *d'Orb. Voy., Ois.* p. 285; *Burm. La-Plata Reise,* ii. p. 485 (Mendoza); *Scl. et Salv. Nomencl.* p. 32; *Salvin, Ibis,* 1880, p. 355; *Döring, Exp. al Rio Negro, Zool.* p. 40 (R. Colorado).

Description.—Similar to *Embernagra platensis*, but the back unstriped, and the olive hue somewhat purer; also the abdomen of a paler buff: total length 8·1 inches, wing 3·7, tail 4·0. *Female* similar.

Hab. Bolivia, Western Argentina, and Patagonia.

This species, which was met with by Dr. Burmeister near Mendoza, was found by Dr. Döring as far south as the Rio Colorado, on the pampas.

---

85. EMBERIZOIDES SPHENURUS (Vieill.).

(WEDGE-TAILED GROUND-FINCH.)

Embernagra macroura, *d'Orb. Voy., Ois.* p. 285. **Emberizoides macrurus,** *Burm. La-Plata Reise,* ii. p. 485 (Paraná). **Emberizoides sphenurus,** *Scl. et Salv. Nomencl.* p. 33.

Description.—Above yellowish brown, striped with black; shoulders edged with yellow; wing-feathers blackish, edged with olive-green; tail-feathers blackish, edged with pale brown; beneath pale ochaceous brown, white on the throat and middle of the belly; bill and feet pale horn-colour: whole length 8·0 inches, wing 3·1, tail 4·0.

Hab. Brazil, Bolivia, Paraguay, and Argentina.

Burmeister met with this species near Paraná and in other parts of Northern Argentina.

---

86. HÆMOPHILA WHITII (Sharpe).

(WHITE'S GROUND-FINCH.)

Zonotrichia whitii, *Sharpe, Cat. B.* xiii. p. 608, pl. xiii. **Zonotrichia strigiceps,** *White, P. Z. S.* 1883, p. 38 (Cordova).

Description.—Above, head and neck grey, variegated with dark chestnut; back yellowish brown with black stripes; wings and tail blackish, the feathers edged with pale brown; beneath, throat and belly white, breast pale grey; sides yellowish brown; bill dark horn-colour, lower mandible whitish; feet light horn-colour: whole length 6·0 inches, wing 2·5, tail 2·8.

Hab. Northern Argentina.

This species has been based by Mr. Sharpe on a specimen, obtained by White near Cordova, which was wrongly determined as *Zonotrichia strigiceps*—a species that it somewhat resembles in its upper plumage.

---

87. CHRYSOMITRIS ICTERICA (Licht.).

(BLACK-HEADED SISKIN.)

Chrysomitris barbata, *Scl. et Salv. P. Z. S.* 1868, p. 140 (Buenos Ayres); *Durnford, Ibis,* 1877, p. 172 (Buenos Ayres); *White, P. Z. S.* 1882, p. 600 (Catamarca, Misiones); *Döring, Exp. al Rio Negro, Zool.* p. 40 (R. Sauce, R. Colorado, R. Negro); *Barrows, Bull. Nutt. Orn. Cl.* viii. p. 132 (Concepcion). **Chrysomitris magellanica**, *Burm. La-Plata Reise,* ii. p. 489; *Gibson, Ibis,* 1880, p. 30 (Buenos Ayres); *Salvin, Ibis,* 1880, p. 355 (Salta). **Chrysomitris icterica**, *Sharpe, Cat. B.* xii. p. 217.

Description.—Above light olive-green; lesser wing-coverts same as the back; wings black, a broad bright yellow band across the base of the feathers; rump yellow; upper tail-coverts olive-green; tail-feathers yellow at the base and black at the ends; head all round and throat velvety black; beneath and under wing-coverts bright yellow: total length 5·0 inches, wing 3·0, tail 1·8. *Female* without the black on the head, otherwise similar to the male, but less bright.

Hab. Brazil and Argentina.

This beautiful little golden-plumaged Finch, the male distinguished from his consort by a brighter yellow colour and a black head, is extremely common throughout the entire length of the Argentine country from Brazil to Patagonia. In the Buenos-Ayrean district it probably has a partial migration, as small flocks are seen to arrive in spring; but further south, in Patagonia, it appears to be strictly resident. In settled districts they are always more abundant than in the woods, and they have a special predilection for poplar groves, and always prefer a poplar to build in. They go in small flocks, seldom more than about a dozen birds together, have a rapid, undulating flight, feed chiefly on the ground like most Finches, and also frequently alight in the seeding-time on plants like the lettuce and *Sonchus asper* (a common weed) and, clinging to the stem, dexterously pick off the seed, scattering the down about them in a little cloud. They are very tuneful, restless, quick in their motions, apparently always in a light-hearted merry mood. Being much admired for their song, they are often kept in cages; and certainly, for cheerfulness and constancy in singing, they take the foremost place amongst the Finches; but there is little expression in the song, which is composed of a variety of short twittering notes, uttered with great rapidity, as the bird sits perched on a twig or undulates from tree to tree. Usually the notes flow in a continuous stream, but occasionally the bird sings in a different manner, making a pause of two or three seconds of silence after every eight or ten short notes. When the female is on the nest the male sometimes perches near her amongst the leaves and sings *sotto voce*, apparently for her pleasure only, the notes being so low that, at a distance of ten yards, they can scarcely be heard.

The nest is usually placed between the angle formed by a small branch and the bole of the tree, and is a deep, well-made structure composed of many materials, and lined with horsehair, down, or feathers. The eggs are five, very small for the bird, pure white, and so frail that it is not easy to take them from the nest without breaking them.

While engaged in building, the birds constantly utter a low, soft, trilling note; and when the nest is approached they break out into long, somewhat reedy notes, resembling those of the Canary, expressive of alarm or curiosity.

---

88. CHRYSOMITRIS ATRATA (d'Orb. et Lafr.).

(HALF-BLACK SISKIN.)

Carduelis atratus, *d'Orb. Voy., Ois.* p. 364, t. 48. fig. 2. **Chrysomitris atrata,** *Burm. La-Plata Reise,* ii. p. 490 (Mendoza); *Sharpe, Cat. B.* xii. p. 212.

Description.—Black; concealed shoulder-spot, broad band on wing, basal half of tail-feathers, belly, and under tail-coverts bright yellow; bill and feet horn-colour: whole length 5·4 inches, wing 2·9, tail 1·8.

Hab. Bolivia and N.W. Argentina.

Dr. Burmeister obtained two examples of this species in the Sierra of Uspellata near Mendoza.

---

89. SYCALIS PELZELNI, Scl.

(YELLOW HOUSE-SPARROW.)

Sycalis pelzelni, *Scl. Ibis,* 1872, p. 42; *Scl. et Salv. Nomencl.* p. 34; *Durnford, Ibis,* 1877, p. 172 (Buenos Ayres).

Description.—Above yellowish olive-green, the back sparsely striped with blackish; wing- and tail-feathers black, edged with yellow; forehead bright orange, the rest of the head like the back; below bright yellow; under surfaces of wings and tail also yellow: total length 5·4 inches, wing 2·7, tail 2·2. *Female* dull brownish grey mottled with blackish above; under surface whitish grey, striped with dusky brown on the breast; wing- and tail-feathers edged with yellow.

Hab. S. Brazil, Paraguay, and Argentina.

The Yellow "House-Sparrow," as this species is called, is the town-bird of Buenos Ayres, but does not multiply greatly, nor is he familiar with man, like his rough, sooty-plumaged, far-away London relation.

The forehead of the male is bright orange, the prevailing colour of the entire plumage yellow, clouded with other hues. The female is grey, marked with pale fuscous, and is less in size than her mate. They remain with us all the year and live in pairs, the sexes in this species being faithful. Sometimes they are seen associating in small flocks, but I am inclined to believe that only the young unmated birds are gregarious. In 1867–8, during the cholera epidemic in Buenos Ayres, the Sparrows all disappeared from the town, and I was told by the manager of a large steam flour-mill in the town that the birds had not gone away, but had died. They were found dead all about the mill where they had been very abundant. My informant was a careful observer, and I have no doubt that he was correct in what he told me.

In spring and summer the male sings frequently with great energy, but without much melody. After a hurried prelude of sharp chirps and trills, he pours out a continuous stream of sound, composed of innumerable brief notes, high and shrill as those of a bat, wounding the ear with their excessive sharpness, and emitted so rapidly that the whole song is more like that of a cicada than of a bird. This piercing torrent of sound is broken at intervals by a long grave note, or half a dozen short rapid notes in a lower key, which come as an agreeable relief.

In towns they build in walls, like the English Sparrow; in country places they always select the domed nest of some Dendrocolaptine species to breed in. Possibly in some districts where I have not been, this Sparrow selects other breeding-sites; my experience is that outside of a town it never lays anywhere but in some domed nest, and at home I frequently put up boxes for them in the trees, but they would not notice them, though the Wrens and Swallows were glad to have them. Sometimes they make choice of the large fabric of the *Anumbius acuticaudatus*, called Leñatero in the vernacular; but their claim to this nest (even when the Leñateros are out of it) is frequently disputed by other species which possess the same habit as this Sparrow, but are more powerful than he. Their favourite breeding-place is, however, the solid earthen structure of the Oven-bird; and it is wonderful to see how persistently and systematically they labour to drive out the lawful owners—birds so much larger and more powerful than themselves. Early in spring, and before the advent of the Tree-Martins, the pair of Sparrows begin haunting the neighbourhood of the oven they have elected to take possession of, usually one pretty high up in a tree. As the season advances their desire towards it increases, and they take up their position on the very tree it is in; and finally a particular branch near the oven, commanding a good view of the entrance, is chosen for a permanent resting-place. Here they spend a great portion of their time in song, twitterings, and loving dalliance, and, if attentively observed, they are seen with eyes ever fixed on the coveted abode. As the need for a receptacle for the eggs becomes more urgent they grow bolder, and in the absence of the owners flit about the oven, alight on it, and even enter it. The Oven-bird appears to drive them off with screams of indignation, but the moment he retires they are about it again, and, even when it contains eggs or young birds, begin impudently carrying in feathers, straws, and other materials for a nest, as if they were already in undisputed possession. At this stage the Tree-Martins (*Progne tapera*) perhaps appear to complicate matters; and even if these last comers do not succeed in ousting the Oven-birds, they are sure to seize the oven when it becomes vacant, and the Sparrows, in spite of their earlier claim, are left out in the cold. But they do not take their defeat quietly, or, rather, they do not know when they are beaten, but still remain to harrass their fellow-pirates, just as they did the Oven-birds before, bringing straws and feathers in their beaks, and when forced to drop these materials and chased from the neighbourhood with great noise and fury by the Tree-Martins, it is only to return undaunted in a few minutes, bringing more straws and feathers.

This Sparrow makes a rather large nest, neatly lined with horsehair, and lays five eggs, long, pointed, the entire surface thickly marked with deep chocolate-brown.

In rural districts this species is comparatively rare, not more than one or two couples being seen about each habitation; and I scarcely think it would be too much to say that there are four or five thousand Chingolos for every individual Yellow Sparrow. Yet it is a hardy little bird, well able to hold its own, subsists on the same kind of food and lays as many eggs as the *Zonotrichia*; and it possesses, moreover, a great advantage over the dominant species in placing its nest out of the reach of the parasitical *Molothrus*, the destroyer of about fifty per cent. of the Chingolo's eggs. I can only attribute the great disparity in the numbers of the two species to the fact that the Yellow House-Sparrow will breed only (out of towns) in nests not easily taken, and to the stubborn pertinacity which leads it to waste the season in these vain efforts, while the other species is rearing its brood. This is a blunder of instinct comparable to that of the Minera (*Geositta cunicularia*), mentioned by Darwin in the 'Voyage of a Naturalist,' where the bird made its hole in a mud wall a few inches wide, and on coming out on the other side simply went back and made another hole, and then another, unable to understand that the wall had not the requisite width.

In such a case as the Yellow House-Sparrow presents, in which the colour of the sexes differs, the female being without any of the brighter hues found in the male, and which makes an elaborate nest and lays deeply-coloured eggs, it is impossible not to believe that the bird originally built in exposed situations, and subsequently—perhaps in very recent times—acquired the habit of breeding in dark holes. The frequent destruction of the exposed nest, and an abundance of vacant domed nests, into which some individuals occasionally penetrated to breed, would lead to the acquisition of such a nesting-habit; for the birds inheriting it would have an advantage and be preserved, while those persisting in the old habit of building exposed nests would perish. Domed nests made by Dendrocolaptine birds are very abundant even now, and it is probable that, before the country became settled by Europeans, they were very much more numerous. Darwin, speaking of the Oven-bird's habit of always placing its oven in the most conspicuous and (to man) accessible places, predicts, and truly I believe, that this habit will eventually cause the extinction of the species; for when the country becomes more thickly settled, the bird-nesting boys will destroy all the ovens. Probably when the Oven-birds were more abundant the Sparrows could always find vacant ovens to breed in, until a habit of breeding almost exclusively in these safe and convenient bird-built houses was acquired; and the present seemingly stupid persistence of the birds in struggling to get possession of those already occupied by stronger species, only shows that the habit or instinct has not been modified to suit a change in the conditions—*i. e.* a diminishing number of ovens to breed in, with, perhaps, the increase of other stronger species possessing the same habit. But while the instinct thus survives too strongly in the country birds, many individuals have taken to a town life, and acquired the new habit of breeding in holes in brick walls. Probably this race of town birds will eventually colonize the rural districts, and usurp the place of the country birds, which will then be placed at a disadvantage.

---

90. SYCALIS LUTEA (d'Orb. et Lafr.).

(YELLOW SEED-FINCH.)

Sycalis chloropis, *Burm. La-Plata Reise,* ii. p. 489 (Mendoza and Catamarca).
Sycalis lutea, *Scl. Ibis,* 1872, p. 46, pl. ii. fig. 2.

Description.—Dark yellow; rump and body below brighter; wings and tail brownish black, edged with yellow; under wing-coverts pale yellow; inner margins of wing-feathers pale brown: whole length 5·5 inches, wing 3·2, tail 2·1. *Female* similar, but duller and more brownish.

Hab. Andes of Peru, Bolivia, and Argentina.

Prof. Burmeister obtained examples of this species near Mendoza. In my revision of the genus published in 1872, I referred the specimens (upon some of which Prof. Burmeister based his species *S. chloropis*) to *S. uropygialis.* I now find that this was an error, and that they really belong to *S. lutea.*—P. L. S.

---

91. SYCALIS LUTEOLA (Sparrm.).

(MISTO SEED-FINCH.)

Sycalis luteola, *Scl. Ibis,* 1872, p. 44; *Scl. et Salv. Nomencl.* p. 35; *Durnford, Ibis,* 1877, p. 172 (Buenos Ayres), et 1878, p. 304 (Centr. Pat.); *Barrows, Bull. Nutt. Orn. Cl.* viii. p. 132 (Entrerios). **Sycalis luteiventris,** *Burm. La-Plata Reise,* ii. p. 489.

Description.—Above light olive-green, marked with dusky stripes; wing-feathers blackish, with pale brownish edges; tail-feathers the same; beneath, throat and chest dusky buff, lower breast and belly yellow; bill and feet horn-colour: total length 5·0 inches, wing 2·6, tail 2·0. *Female* similar, but not so bright.

Hab. S. America, from Colombia to Chili.

This is a slender, graceful bird, less than the Canary in size, the whole upper plumage yellowish olive, with dun markings, the lower surface of a dull yellow. The female is a little smaller than the male, and her colours are somewhat dimmer.

This species is resident and gregarious in the Argentine Republic, and in autumn frequently congregates in flocks of several thousands. They are not so universally distributed as the Chingolo, and are not wood-birds, but frequent open plains abounding in thistles and other coarse herbage, which affords them shelter. In cultivated districts, where their food is most abundant, they are excessively numerous, and, after the harvest has been gathered, frequent the fields in immense flocks. While feeding, the flocks scatter over a large area of ground, being broken up into small companies of a dozen or more birds, and at such times are so intent on their food that a person can walk about amongst them without disturbing them. They take flight very suddenly, bursting into a thousand chirping, scolding notes, pursue each other through the air, and, after wheeling about the field for a minute or two, suddenly drop down into the grass again and are silent as before.

In August they begin to sing, here and there an individual being heard in the fields; but when the weather grows warmer they repair to the plantations in vast numbers, and, sitting on the branches, sing in a concert of innumerable voices, which produces a great volume of confused sound, and which often continues for hours at a time without intermission.

By-and-by these pleasant choirs break up, the birds all scattering over the plains and fields to woo and build, and it is then first discovered that the male has a peculiar and very sweet song. Apart from his fellows, he acquires a different manner of singing, soaring up from his stand on the summit of a bush or stalk, and beginning his song the moment he quits his perch. Ascending, he utters a series of long, melodious notes, not loud, but very distinctly enunciated and increasing in volume; at a height of fifty or sixty yards he pauses, the notes becoming slower; then, as he descends with a graceful spiral flight, the wings outstretched and motionless, the notes also fall, becoming lower, sweeter, and more impressive till he reaches the earth. After alighting the song continues, the notes growing longer, thinner, and clearer, until they dwindle to the merest threads of sound, and cease to be audible except to a person standing within a few yards of the songster. The song is quite unique in character, and its great charm is in its gradual progress from the somewhat thick notes at the commencement to the thin, tremulous tones with which the bird returns to earth, and which change again to the excessively attenuated sounds at the end.

The nest is deep, well-built, and well-concealed, sometimes resting on the ground, but frequently raised above it. It contains five long, pointed eggs, with a white or bluish-white ground-colour, and thickly spotted with brown. I have frequently found the eggs of the *Molothrus* in its nest, but have never been able to see this Sparrow feeding, or followed by, a young *Molothrus*. Possibly, if it ever hatches the parasitical egg at all, the voracious young Cow-bird is starved by the delicate food supplied by its foster-parents.

---

92. OROSPINA PRATENSIS, Cab.

(MEADOW SEED-FINCH.)

Orospina pratensis, *Cab. Journ. f. Orn.* 1883, p. 108, tab. i. fig. 1.

Description.—Cap dark greenish yellow; rump yellowish green; remainder of upper parts dark brown, feathers of interscapulium and of wings and tail edged with yellowish green; inner webs of the outer pair of tail-feathers almost wholly white, the next pair with a large white spot; under surface yellow, lightest on the throat, middle of belly, and crissum; flanks greenish: total length 4·5 inches, wing 2·5, tail 1·8.

Hab. Tucuman.

Herr Schulz discovered this little Finch, which Dr. Cabanis has referred to a new genus allied to *Sycalis*, upon the high Sierras of Tucuman, where it was observed sitting on the stones.

Fam. XI. ICTERIDÆ, OR TROUPIALS.

The Icteridæ or Troupials constitute a large group of Passerine birds allied to our Starlings (*Sturnidæ*), of which they take the place in the New World. They are at once structurally distinguishable from the Starlings by having only 9 primaries in the wing, just as the Mniotiltidæ of the New World are in a similar manner distinguishable from the Sylviidæ.

In America the Icteridæ play an important part, numbering some 130 species, and extending throughout the two continents from north to south. Of these, 15 species occur in Argentina, and amongst them are three species of Cow-bird (*Molothrus*), remarkable for their parasitic habits, of which Hudson's observations have enabled him to give a full and, for the first time, a tolerably complete account.

---

93. AMBLYCERCUS SOLITARIUS (Vieill.).

(SOLITARY CASSIQUE.)

Cassicus solitarius, *Scl. et Salv. Nomencl.* p. 36; *Barrows, Bul. Nutt. Orn. Cl.* viii. p. 133 (Entrerios); *Scl. Cat. B.* xi. p. 326.

Description.—Uniform black; bill white; feet black: total length 11 inches, wing 4·8, tail 4·5. *Female* similar, but smaller.

Hab. Peru, Bolivia, Brazil, Paraguay, and Northern Argentina.

Mr. Barrows obtained a single specimen of this species at Concepcion, and others were seen. It was said to be an excellent song-bird, and to be more abundant further up the Uruguay River.

---

94. MOLOTHRUS BONARIENSIS (Gm.).

(ARGENTINE COW-BIRD.)

Molothrus bonariensis, *Scl. et Salv. Nomencl.* p. 37; *Hudson, P. Z. S.* 1872, p. 809, 1874, p. 153 (Buenos Ayres); *Durnford, Ibis,* 1877, pp. 33, 174 (Chupat); *White, P. Z. S.* 1882, p. 601 (Buenos Ayres); *Döring, Exp. al Rio Negro, Zool.* p. 41 (Carhue); *Barrows, Bull. Nutt. Orn. Cl.* viii. p. 133 (Entrerios); *Scl. Cat. B.* xi. p. 335. **Molothrus sericeus,** *Burm. La-Plata Reise,* ii. p. 494.

Description.—Uniform shining purplish black; less lustrous on wings and tail; bill and feet black: total length 7·5 inches, wing 4·5, tail 3·0. *Female* dark ashy brown, beneath paler; slightly smaller in size.

Hab. Argentina, Paraguay, Bolivia, and Brazil.

This species is the *Tordo Comun* of Azara, and is usually called "*Tordo*" or "*Pajaro Negro*" by the Spanish, and "*Blackbird*" by the English-speaking Argentines. A more suitable name, I think, is the Argentine Cow-bird, which has been given to it by some writers on ornithology, Cow-bird being the name of the closely allied North-American species, *Molothrus pecoris.*

This Cow-bird is widely distributed in South America, and is common throughout the Argentine country, including Patagonia, as far south as Chupat. In Buenos Ayres it is very numerous, especially in cultivated districts where there are plantations of trees. The male is clothed in a glossy plumage of deep violaceous purple, the wings and tail being dark metallic green; but seen at a distance or in the shade the bird looks black. The female is inferior in size and has a dull, mouse-coloured plumage, and black beak and legs. The males are much more numerous than the females. Azara says that nine birds in ten are males; but I am not sure that the disparity is so great as that. It seems strange and contrary to Nature's usual rule that the smaller, shyer, inconspicuous individuals should be in such a minority; but the reason is perhaps that the *male eggs* of the Cow-bird are harder-shelled than the *female eggs*, and escape destruction oftener, when the parent bird exercises its disorderly and destructive habit of pecking holes in all the eggs it finds in the nests into which it intrudes.

The Cow-birds are sociable to a greater degree than most species, their companies not breaking up during the laying-season; for, as they are parasitical, the female merely steals away to drop her egg in any nest she can find, after which she returns to the flock. They feed on the ground, where in their movements and in the habit the male has of craning out its neck when disturbed, they resemble Starlings. The male has also a curious habit of carrying his tail raised vertically while feeding. They follow the domestic cattle about the pastures, and frequently a dozen or more birds may be seen perched along the back of a cow or horse. When the animal is grazing they group themselves close to its mouth, like chickens round a hen when she scratches up the ground, eager to snatch up the small insects exposed where the grass is cropped close. In spring they also follow the plough to pick up worms and grubs.

The song of the male, particularly when making love, is accompanied with gestures and actions somewhat like those of the domestic Pigeon. He swells himself out, beating the ground with his wings, and uttering a series of deep internal notes, followed by others loud and clear; and occasionally, when uttering them, he suddenly takes wing and flies directly away from the female to a distance of fifty yards, and performs a wide circuit about her in the air, singing all the time. The homely object of his short-lived passion always appears utterly indifferent to this curious and pretty performance; yet she must be even more impressionable than most female birds, since she continues scattering about her parasitical and often wasted eggs during four months in every year. Her language consists of a long note with a spluttering sound, to express alarm or curiosity, and she occasionally chatters in a low tone as if trying to sing. In the evening, when the birds congregate on the trees to roost they often continue singing in concert until it is quite dark; and when disturbed at night the males frequently utter their song while taking flight, reminding one of the *Icterus pyrrhopterus*, which has only its usual melody to express fear and other painful emotions. On rainy days, when they are driven to the shelter of trees, they will often sing together for hours without intermission, the blending of innumerable voices producing a rushing sound as of a high wind. At the end of summer they congregate in flocks of tens of thousands, so that the ground where they are feeding seems carpeted with black, and the trees when they alight appear to have a black foliage. At such times one wonders that many small species on which they are parasites do not become extinct by means of their pernicious habit. In Buenos Ayres, where they are most numerous, they have a migration, which is only partial, however. It is noticeable chiefly in the autumn, and varies greatly in different years. In some seasons it is very marked, when for many days in February and March the birds are seen travelling northwards, flock succeeding flock all day long, passing by with a swift low undulating flight, their wings producing a soft musical sound; and this humming flight of the migrating Cow-birds is as familiar to every one acquainted with nature in Buenos Ayres as the whistling of the wind or the distant lowing of cattle.

The procreant instinct of this *Molothrus* has always seemed so important to me, for many reasons, that I have paid a great deal of attention to it; and the facts, or, at all events, the most salient of them, which I have collected during several years of observation, I propose to append here, classified under different headings so as to avoid confusion and to make it easy for other observers to see at a glance just how much I have learnt.

Though I have been familiar with this species from childhood, when I used to hunt every day for their wasted eggs on the broad, clean walks of the plantation, and removed them in pity from the nests of little birds where I found them, I have never ceased to wonder at their strange instinct, which in its wasteful destructive character, so unlike the parasitical habit in other species, seems to strike a discordant note in the midst of the general harmony of nature.

Mistakes and Imperfections of the Procreant Instinct of Molothrus bonariensis.

1. The Cow-birds, as we have seen, frequently waste their eggs by dropping them on the ground.

2. They also occasionally lay in old forsaken nests. This I have often observed, and to make very sure I took several old nests and placed them in trees and bushes, and found that eggs were laid in them.

3. They also frequently lay in nests where incubation has actually begun. When this happens the Cow-bird's egg is lost if incubation is far advanced; but if the eggs have been sat on three or four days only, then it has a good chance of being hatched and the young bird reared along with its foster-brothers.

4. One female often lays several eggs in the same nest, instead of laying only one, as does, according to Wilson, the *Molothrus pecoris*

of North America. I conclude that this is so from the fact that in cases where the eggs of a species vary considerably in form, size, and markings, each individual of the species lays eggs precisely or nearly alike. So when I find two, three, or four eggs of the Cow-bird in one nest all alike in colour and other particulars, and yet in half a hundred eggs from other nests cannot find one to match with them, it is impossible not to believe that the eggs found together, and possessing a family likeness, were laid by the same bird.

5. Several females often lay in one nest, so that the number of eggs in it frequently makes incubation impossible. One December I collected ten nests of the Scissor-tail (*Milvulus tyrannus*) from my trees; they contained a total of 47 eggs, 12 of the Scissor-tails and 35 of the Cow-birds. It is worthy of remark that the *Milvulus* breeds in October or early in November, rearing only one brood; so that these ten nests found late in December were of birds that had lost their first nests. Probably three fourths of the lost nests of *Milvulus* are abandoned in consequence of the confusion caused in them by the Cow-birds.

6. The Cow-birds, male and female, destroy many of the eggs in the nests they visit, by pecking holes in the shells, breaking, devouring, and stealing them. This is the most destructive habit of the bird, and is probably possessed by individuals in different degrees. I have often carefully examined all the parasitical eggs in a nest, and after three or four days found that these eggs had disappeared, others, newly laid, being in their places. I have seen the female Cow-bird strike her beak into an egg and fly away with it; and I have often watched the male bird perched close by while the female was on the nest, and when she quitted it seen him drop down and begin pecking holes in the eggs. In some nests found full of parasitical eggs every egg has holes pecked in the shell, for the bird destroys indiscriminately eggs of its own and of other species.

Advantages possessed by M. bonariensis *over its dupes.*

After reading the preceding notes one might ask, If there is so much that is defective and irregular in the reproductive instinct of *M. bonariensis*, how does the species maintain its existence, and even increase to such an amazing extent? for it certainly is very much more numerous, over an equal area, than other parasitical species. For its greater abundance there may be many reasons unknown to us. The rarer species may be less hardy, have more enemies, be exposed to more perils in their long migrations, &c. That it is able to maintain its existence in spite of irregularities in its instinct is no doubt due to the fact that its eggs and young possess many advantages over the eggs and young of the species upon which it is parasitical. Some of these advantages are due to those very habits of the parent bird which at first sight appear most defective; others to the character of the egg and embryo, time of evolution, &c.

1. The egg of the Cow-bird is usually larger, and almost invariably harder-shelled than are the eggs it is placed with; those of the Yellow-breast (*Pseudoleistes virescens*) being the one exception I am acquainted with. The harder shell of its own egg, considered in relation to the destructive egg-breaking habit of the bird, gives it the best chance of being preserved; for though the Cow-bird never distinguishes its own eggs, of which indeed it destroys a great many, a larger proportion escape in a nest where many eggs are indiscriminately broken.

2. The vitality or tenacity of life appears greater in the embryo Cow-bird than in other species; this circumstance also, in relation to the egg-breaking habit and to the habit of laying many eggs in a nest, gives it a further advantage. I have examined nests of the Scissor-tail, containing many eggs, after incubation had begun, and have been surprised at finding those of the Scissor-tail addled, even when placed most advantageously in the nest for receiving heat from the parent bird, while those of the Cow-bird contained living embryos, even when under all the other eggs, and, as frequently happens, glued immovably to the nest by the matter from broken eggs spilt over them.

The following instance of extraordinary vitality in an embryo *Molothrus* seems to show incidentally that in some species protective habits, which will act as a check on the parasitical instinct, may be in the course of formation.

Though birds do not, as a rule, seem able to distinguish parasitical eggs from their own, however different in size and colour they may be, they often do seem to know that eggs dropped in their nest before they themselves have began to lay ought not to be there; and the nest, even after its completion, is not infrequently abandoned on account of these premature eggs. Some species, however, do not forsake their nests; and though they do not throw the parasitical eggs out, which would seem the simplest plan, they have discovered how to get rid of them and so save themselves the labour of making a fresh nest. Their method is to add a new deep lining, under which the strange eggs are buried out of sight and give no more trouble. The *Sisopygis icterophrys*—a common Tyrant-bird in Buenos Ayres—frequently has recourse to this expedient; and the nest it makes being rather shallow the layer of fresh material, under which the strange eggs are buried, is built upwards above the rim of the original nest; so that this supplementary nest is like one saucer placed within another, and the observer is generally able to tell from the thickness of the whole structure whether any parasitical eggs have been entombed in it or not. Finding a very thick nest one day, containing two half-fledged young birds besides three addled eggs, I opened it, removing the upper portion, or additional nest, intact, and discovered beneath it three buried *Molothrus* eggs, their shells encrusted with dirt and glued together with broken egg-matter spilt over them. In trying to get them out without pulling the nest to pieces I broke them all; two were quite rotten, but the third contained a living embryo, ready to be hatched, and very lively and hungry when I took it in my hand. The young Tyrant-birds were about a fortnight old, and as they hatch out only about twenty days after the parent-bird begins laying, this parasitical egg with a living chick in it must have been deeply buried in the nest for five or six weeks. Probably after the young Tyrant-birds came out of their shells and began to grow, the little heat from their bodies penetrating to the buried egg, served to bring the embryo in it to maturity; but when I saw it I felt (like a person who sees a ghost) strongly inclined to doubt the evidence of my own senses.

3. The comparatively short time the embryo takes to hatch gives it another and a great advantage; for, whereas the eggs of other small birds require from fourteen to sixteen days to mature, that of the Cow-bird hatches in eleven days and a half from the moment incubation commences; so that when the female Cow-bird makes so great a mistake as to drop an egg with others that have already been sat on, unless incubation be very far advanced, it still has a chance of being hatched before or contemporaneously with the others; but even if the others hatch first, the extreme hardiness of the embryo serves to keep it alive with the modicum of heat it receives.

4. Whenever the *Molothrus* is hatched together with the young of its foster-parents, if these are smaller than the parasite, as usually is the case, soon after exclusion from the shell they disappear, and the young Cow-bird remains sole occupant of the nest. How it succeeds in expelling or destroying them, if it indeed does destroy them, I have not been able to learn.

5. To all these circumstances favourable to the *Molothrus* may be added another of equal or even greater importance. It is never engaged with the dilatory and exhaustive process of rearing its own young; and for this reason continues in better condition than other species, and, moreover, being gregarious and practising promiscuous sexual intercourse, must lay a much greater number of eggs than other species. In our domestic fowls we see that hens that never become broody lay a great deal more than others. Some of our small birds rear two, others

only one brood in the season—building, incubation, and tending the young taking up much time, so that they are usually from two to three months and a half employed. But the Cow-bird is like the fowl that never incubates, and continues dropping eggs during four months and a half. From the beginning of September until the end of January the males are seen incessantly wooing the females, and during most of this time eggs are found. I find that small birds will, if deprived repeatedly of their nests, lay and even hatch four times in the season, thus laying, if the full complement be four, sixteen eggs. No doubt the Cow-bird lays a much larger number than that; my belief is that every female lays from sixty to a hundred eggs every season, though I have nothing but the extraordinary number of wasted eggs one finds to judge from.

Before dismissing the subject of the advantages the *Molothrus* possesses over its dupes, and of the real or apparent defects of its instinct, some attention should be given to another circumstance, viz., the new conditions introduced by land-cultivation and their effect on the species. The altered conditions have, in various ways, served to remove many extraneous checks on the parasitical instinct, and the more the birds multiply, the more irregular and disordered does the instinct necessarily become. In wild districts where it was formed, and where birds building accessible nests are proportionately fewer, the instinct seems different from what it does in cultivated districts. Parasitical eggs are not common in the desert, and even the most exposed nests there are probably never overburdened with them. But in cultivated places, where their food abounds, the birds congregate in the orchards and plantations in great numbers, and avail themselves of all the nests, ill-concealed as they must always be in the clean, open-foliaged trees planted by man.

Diversity in Colour of Eggs.

There is an extraordinary diversity in the colour, form, and disposition of markings &c. of the eggs of *M. bonariensis*; and I doubt whether any other species exists laying eggs so varied. About half the eggs one finds, or nearly half, are pure unspotted white, like the eggs of birds that breed in dark holes. Others are sparsely sprinkled with such exceedingly minute specks of pale pink or grey, as to appear quite spotless until closely examined. After the pure white, the most common variety is an egg with a white ground, densely and uniformly spotted or blotched with red. Another not uncommon variety has a very pale, flesh-coloured ground, uniformly marked with fine characters, that look as if inscribed on the shell with a pen. A much rarer variety has a pure white shell with a few large or variously sized chocolate spots.

Perhaps the rarest variety is an egg entirely of a fine deep red; but between this lovely marbled egg and the white one with almost imperceptible specks, there are varieties without number; for there is no such thing as characteristic markings in the eggs of this species, although, as I have said before, the eggs of the same individual show a family resemblance.

Habits of the young M. bonariensis.

Small birds of all species, when first hatched, closely resemble each other; after they are fledged the resemblance is less, but still comparatively great; grey, interspersed with brown, is the colour of most of them, or at least of the upper exposed plumage. There is also a great similarity in their cries of hunger and fear—shrill, querulous, prolonged, and usually tremulous notes. It is not, then, to be wondered at that the foster-parents of the young *Molothrus* so readily respond to its cries, understanding the various expressions denoting hunger, fear, pain, as well as when uttered by their own offspring. But the young *Molothrus* never understands the language of its foster-parents as other young birds understand the language of their real parents, rising to receive food when summoned, and concealing themselves or trying to escape when the warning note is given. How does the young *Molothrus* learn to distinguish, even by sight, its foster-parent from any other bird approaching the nest? It generally manifests no fear even at a large object. On thrusting my fingers into any nest, I find the young birds, if still blind or but recently hatched, will hold up and open their mouths expecting food; but in a very few days they learn to distinguish between their parents and other objects approaching them, and to show alarm even when not warned of danger. Consider the different behaviour of three species that seldom or never warn their offspring of danger. The young of *Synallaxis spixi,* though in a deep domed nest, will throw itself to the ground, attempting thus to make its escape. The young of *Mimus patagonicus* sits close and motionless, with closed eyes, mimicking death. The young of our common *Zenaida,* even before it is fledged, will swell itself up and strike angrily at the intruder with beak and wings; and, by making so brave a show of its inefficient weapons, it probably often saves itself from destruction. But any thing approaching the young *Molothrus* is welcomed with fluttering wings and clamorous cries, as if all creatures were expected to minister to its necessities.

December 24.—To-day I found a young *Molothrus* in the nest of *Spermophila cærulescens*; he cried for food on seeing my hand approach the nest; I took him out and dropped him down, when, finding himself on the ground, he immediately made off, half-flying. After a hard chase I succeeded in recapturing him, and began to twirl him about, making him scream, so as to inform his foster-parents of his situation, for they were not by at the moment. I then put him back in, or rather upon, the little cradle of a nest, and plucked half-a-dozen large measure-worms from an adjacent twig. The worms I handed to the bird as I drew them from the cases, and with great greediness he devoured them all, notwithstanding the ill-treatment he had just received, and utterly disregarding the wild excited cries of his foster-parents, just arrived and hovering within three or four feet of the nest.

Last summer I noticed a young Cow-bird in a stubble-field, perched on the top of a slender dry stalk; as it was clamouring at short intervals, I waited to see what bird would come to it. It proved to be the diminutive *Hapalocercus flaviventris*; and I was much amused to see the little thing fly directly to its large foster-offspring and, alighting on its back, drop a worm into the upturned open mouth. After remaining a moment on its singular perch, the Flycatcher flew away, but in less than half a minute returned and perched again on the young bird's back. I continued watching them until the *Molothrus* flew off, but not before I had seen him fed seven or eight times in the same manner.

In the two foregoing anecdotes may be seen the peculiar habits of the young *Molothrus*. As the nests in which it is hatched, from those of the little *Serpophaga* and Wren to those of *Mimus*, vary so much in size and materials, and are placed in such different situations, the young *Molothrus* must have in most of them a somewhat incongruous appearance. But in the habits of the young bird is the greatest incongruity or inadaptation. When the nest is in a close thicket or forest, though much too small for the bird, and although the bird itself cannot understand its foster-parents, and welcomes all things that, whether with good or evil design, come near it, the unfitness is not so apparent as when the nest is in open fields and plains.

The young *Molothrus* differs from the true offspring of its foster-parents in its habit of quitting the nest as soon as it is able, trying to follow the old bird, and placing itself in the most conspicuous place it can find, such as the summit of a stalk or weed, and there demanding food with frequent and importunate cries. Thus the little Flycatcher had acquired the habit of perching on the back of its charge to feed it, because parent birds invariably perch above their young to feed them, and the young Cow-bird prevented this by always sitting on the summit of the stalk it perched on. The habit is most fatal on the open and closely cropped pampas inhabited by the Cachila (*Anthus correndera*).

In December, when the Cachila Pipit rears its second brood, the *Milvago chimango* also has young, and feeds them almost exclusively on the young of various species of small birds. At this season the Chimango destroys great numbers of the young of the Cachila and of *Synallaxis hudsoni*. Yet these birds are beautifully adapted in structure, coloration, and habits to their station. It thus happens that in districts where the *Molothrus* is abundant, their eggs are found in a majority of the Cachilas' nests: and yet to find a young Cow-bird out of the nest is a rare thing here, for as soon as the young birds are able to quit the nest and expose themselves they are all or nearly all carried off by the Chimangos.

Conjectures as to the Origin of the Parasitic Instinct in M. bonariensis.

Darwin's opinion that the "immediate and final cause of the Cuckoo's instinct is that she lays her eggs not daily, but at intervals of two or three days" ('Origin of Species'), carries no great appearance of probability with it; for might it not just as reasonably be said that the parasitic instinct is the immediate and final cause of her laying her eggs at long intervals? If it is favourable to a species with the instinct of the Cuckoo (and it probably is favourable) to lay eggs at longer intervals than other species, then natural selection would avail itself of every modification in the reproductive organs that tended to produce such a result, and make the improved structure permanent. It is said ('Origin of Species,' chapter vii.) that the American Cuckoo lays also at long intervals, and has eggs and young at the same time in its nest, a circumstance manifestly disadvantageous. Of the *Coccyzus melanocoryphus*, the only one of our three *Coccyzi* whose nesting-habits I am acquainted with, I can say that it never begins to incubate till the full complement of eggs are laid—that its young are hatched simultaneously. But if it is sought to trace the origin of the European Cuckoo's instinct in the nesting-habits of American *Coccyzi*, it might be attributed not to the aberrant habit of perhaps a single species, but to another and more disadvantageous habit common to the entire genus, viz., their habit of building exceedingly frail platform-nests from which the eggs and young very frequently fall. By occasionally dropping an egg in the deep, secure nest of some other bird, an advantage would be possessed by the birds hatched in them, and in them the habit would perhaps become hereditary. Be this as it may (and the one guess is perhaps as wide of the truth as the other), there are many genera intermediate between *Cuculus* and *Molothrus* in which no trace of a parasitic habit appears; and it seems more than probable that the analogous instincts originated in different ways in the two genera. As regards the origin of the instinct in *Molothrus,* it will perhaps seem premature to found speculations on the few facts here recorded, and before we are acquainted with the habits of other members of the genus. That a species should totally lose so universal an instinct as the maternal one, and yet avail itself of that affection in other species to propagate itself, seems a great mystery. Nevertheless, I cannot refrain from all conjecture on the subject, and will go so far as to suggest what may have been at least one of the many concurrent causes that have produced the parasitic instinct. The apparently transitional nesting-habits of several species, and one remarkable habit of *M. bonariensis,* seem to me to throw some light on a point bearing intimately on the subject, viz., the loss of the nest-making instinct in this species.

Habits vary greatly; were it not so, they would never seem so well adapted to the conditions of life as we find them, since the conditions themselves are not unchangeable. Thus it happens that, while a species seems well adapted to its state in its habits, it frequently seems not so well adapted in its relatively immutable structure. For example, without going away from the pampas, we find a Tringa with the habits of an upland Plover, a Tyrant-bird (*Pitangus bellicosus*) preying on mice and snakes, another Tyrant-bird (*Myiotheretes rufiventris*) Plover-like in its habits, and finally a Woodpecker (*Colaptes campestris*) that seeks its food on the ground like a Starling; yet in none of these—and the list might be greatly lengthened—has there been anything like a modification of structure to keep pace with the altered manner of life. But, however much the original or generic habits of a species may have become altered—the habits of a species being widely different from those of its congeners, also a want of correspondence between structure and habits (the last being always more suited to conditions than the first) being taken as evidence of such alteration—traces of ancient and disused habits frequently reappear. Seemingly capricious actions too numerous, too vague, or too insignificant to be recorded, improvised definite actions that are not habitual, apparent imitations of the actions of other species, a perpetual inclination to attempt something that is never attempted, and attempts to do that which is never done—these and other like motions are, I believe, in many cases to be attributed to the faint promptings of obsolete instincts. To the same cause many of the occasional aberrant habits of individuals may possibly be due—such as of a bird that builds in trees occasionally laying on the ground. If recurrence to an ancestral type be traceable in structure, coloration, language, it is reasonable to expect something analogous to occur in instincts. But even if such casual and often aimless motions as I have mentioned should guide us unerringly to the knowledge of the old and disused instincts of a species, this knowledge of itself would not enable us to discover the origin of present ones. But assuming it as a fact that the conditions of existence, and the changes going on in them, are in every case the fundamental cause of alterations in habits, I believe that in many cases a knowledge of the disused instincts will assist us very materially in the inquiry. I will illustrate my meaning with a suppositititious case. Should all or many species of *Columbidæ* manifest an inclination for haunting rocks and banks, and for entering or peering into holes in them, such vague and purposeless actions, connected with the facts that all Doves build simple platform-nests (like *Columba livia* and others that build on a flat surface), also lay white eggs (the rule being that eggs laid in dark holes are white, exposed eggs coloured), also that one species, *C. livia,* does lay in holes in rocks, would lead us to believe that the habit of this species was once common to the genus. We should conclude that an insufficiency of proper breeding-places, *i. e.* new external conditions, first induced Doves to build in trees. Thus *C. livia* also builds in trees where there are no rocks; but, when able, returns to its ancestral habits. In the other species we should believe the primitive habit to be totally lost from disuse, or only to manifest itself in a faint uncertain manner.

Now, in *Molothrus bonariensis* we see just such a vague, purposeless habit as the imaginary one I have described. Before and during the breeding-season the females, sometimes accompanied by the males, are seen continually haunting and examining the domed nests of some of the Dendrocolaptidæ. This does not seem like a mere freak of curiosity, but their persistence in their investigations is precisely like that of birds that habitually make choice of such breeding-places. It is surprising that they never do actually lay in such nests, except when the side or dome has been accidentally broken enough to admit the light into the interior. Whenever I set boxes up in my trees, the female Cow-birds were the first to visit them. Sometimes one will spend half a day loitering about and inspecting a box, repeatedly climbing round and over it, and always ending at the entrance, into which she peers curiously, and when about to enter starting back, as if scared at the obscurity within. But after retiring a little space she will return again and again, as if fascinated with the comfort and security of such an abode. It is amusing to see how pertinaciously they hang about the ovens of the Oven-birds, apparently determined to take possession of them, flying back after a hundred repulses, and yet not entering them even when they have the opportunity. Sometimes one is seen following a Wren or a Swallow to its nest beneath the eaves, and then clinging to the wall beneath the hole into which it disappeared. I could fill many pages with instances of this habit of *M. bonariensis*, which, useless though it be, is as strong an affection as the bird possesses. That it is a recurrence to a long disused habit, I can scarcely doubt; at least, to no other cause that I can imagine can it be attributed; and, besides, it seems to me that if *M. bonariensis*, when once a nest-builder, had acquired the semiparasitical habit of breeding in domed nests of other birds, such a habit might conduce to the formation of the instinct which it now possesses. I may mention that twice I have seen birds of this species attempting to build nests, and that on both occasions they failed to complete the work. So universal is the nest-making instinct, that one might safely say the *M. bonariensis* had once possessed it, and that in the cases I have mentioned it was a recurrence, too weak to be efficient, to the ancestral habit. Another interesting circumstance may be adduced as strong presumptive evidence that *M. bonariensis* once made itself an open exposed nest as *M. badius* occasionally does—viz., the difference in colour of the male and female; for whilst the former is rich purple, the latter possesses an adaptive resemblance in colour to nests and to the shaded interior twigs and branches on which nests are usually built. How could such an instinct have been lost? To say that the Cow-bird occasionally dropped an egg in another bird's nest, and that the young hatched from these accidental eggs possessed some (hypothetical) advantage over those hatched in the usual way, and that the parasitical habit so became hereditary, supplanting the original one, is an assertion without any thing to support it, and seems to exclude the agency of external conditions. Again the want of correspondence in the habits of the young parasite and its foster-parents would in reality be a disadvantage to the former; the unfitness would be as great in the eggs and other circumstances; for all the advantages the parasite actually possesses in the comparative hardness of the egg-shell, rapid evolution of the young, &c., already mentioned, must have been acquired little by little through the slowly accumulating process of natural selection, but subsequently to the formation of the original parasitical inclination and habit. I am inclined to believe that *M. bonariensis* lost the nest-making instinct by acquiring that semiparasitical habit, common to so many South-American birds, of breeding in the large covered nests of the Dendrocolaptidæ. We have evidence that this semiparasitical habit does tend to eradicate the nest-making one. The *Synallaxes* build great elaborate domed nests, yet we have one species (*S. ægithaloides*) that never builds for itself, but breeds in the nests of other birds of the same genus. In some species the nesting-habit is in a transitional state. *Machetornis rixosa* sometimes makes an elaborate nest in the angle formed by twigs and the bough of a tree, but prefers, and almost invariably makes choice of, the covered nest of some other species or of a hole in the tree. It is precisely the same with our Wren, *Troglodytes furvus*. The Yellow House-Sparrow (*Sycalis pelzelni*) invariably breeds in a dark hole or covered nest. The fact that these three species lay coloured eggs, and the first and last very darkly coloured eggs, inclines one to believe that they once invariably built exposed nests, as *M. rixosa* still occasionally does. It may be added that those species that lay coloured eggs in dark places construct and line their nests far more neatly than do the species that breed in such places but lay white eggs. As with *M. rixosa* and the Wren, so it is with the Bay-winged *Molothrus*; it lays mottled eggs, and occasionally builds a neat exposed nest; yet so great is the partiality it has acquired for large domed nests, that whenever it can possess itself of one by dint of fighting, it will not build one for itself. Let us suppose that the Cow-bird also once acquired the habit of breeding in domed nests, and that through this habit its original nest-making instinct was completely eradicated, it is not difficult to imagine how in its turn this instinct was also lost. A diminution in the number of birds that built domed nests, or an increase in the number of species and individuals that breed in such nests, would involve *M. bonariensis* in a struggle for nests, in which it would probably be defeated. In Buenos Ayres the White-rumped Swallow, the Wren, and the Yellow Seed-finch prefer the ovens of the *Furnarius* to any other breeding-place, but to obtain them are obliged to struggle with *Progne tapera*; for this species has acquired the habit of breeding exclusively in the ovens. They cannot, however, compete with the *Progne*; and thus the increase of one species has, to a great extent, deprived three other species of their favourite building-place. Again, *Machetornis rixosa* prefers the great nest of the *Anumbius*; and when other species compete with it for the nest they are invariably defeated. I have seen a pair of *Machetornis* after they had seized a nest attacked in their turn by a flock of six or eight Bay-wings; but, in spite of the superior numbers, the fury of the *Machetornis* compelled them to raise the siege.

Thus some events in the history of our common *Molothrus* have perhaps been accounted for, if not the most essential one—the loss of the nest-making instinct from the acquisition of the habit of breeding in the covered nests of other birds, a habit that has left a strong trace in the manners of the species, and perhaps in the pure white unmarked eggs of so many individuals; finally, we have seen how this habit may also have been lost. But the parasitical habit of the *M. bonariensis* may have originated when the bird was still a nest-builder. The origin of the instinct may have been in the occasional habit, common to so many species, of two or more females laying together; the progenitors of all the species of *Molothrus* may have been early infected with this habit, and inherited with it a facility for acquiring their present one. *M. pecoris* and *M. bonariensis*, though their instincts differ, are both parasitic on a great number of species; *M. rufoaxillaris* on *M. badius*; and in this last species two or more females frequently lay together. If we suppose that the *M. bonariensis*, when it was a nest-builder, or reared its own young in the nests it seized, possessed this habit of two or more females frequently laying together, the young of those birds that oftenest abandoned their eggs to the care of another would probably inherit a weakened maternal instinct. The continual intercrossing of individuals with weaker and stronger instincts would prevent the formation of two races differing in habit; but the whole race would degenerate, and would only be saved from final extinction by some individuals occasionally dropping their eggs in the nests of other species, perhaps of a *Molothrus*, as *M. rufoaxillaris* still does, rather than of birds of other genera. Certainly in this way the parasitic instinct may have originated in *M. bonariensis* without that species ever having acquired the habit of breeding in the covered dark nests of other birds. I have supposed that they once possessed it only to account for the strange attraction such nests have for them, which seems like a recurrence to an ancestral habit.

---

95. MOLOTHRUS RUFOAXILLARIS, Cassin.

(SCREAMING COW-BIRD.)

[Plate VI. Fig. 2.]

Molothrus rufoaxillaris, *Scl. et Salv. Nomencl.* p. 37; *Hudson, P. Z. S.* 1874, p. 161 (Beunos Ayres); *Durnford, Ibis*, 1877, p. 174 (Buenos Ayres); *White, P. Z. S.* 1882, p. 601 (Catamarca); *Barrows, Bull. Nutt. Orn. Cl.* viii. p. 134 (Entrerios); *Scl. Cat. B.* xi. p. 338.

Description.—Silky black, washed with purple; wings and tail with a slight greenish gloss; a chestnut spot on the axillaries; bill and feet black: whole length 8·0 inches, wing 4·5, tail 3·3. *Female* similar, but somewhat smaller.

Hab. Argentina and Uruguay.

This bird has no vulgar name, not being distinguished from the Common Cow-bird by the country people. The English name of Screaming Cow-bird, which I have bestowed on it, will, I think, commend itself as appropriate to those who observe this bird, for they will always and at any distance be able to distinguish it from the species it resembles so nearly by listening to its impetuous screaming notes, so unlike anything in the language of the Common Cow-bird.

Fig. 1. MOLOTHRUS BADIUS. ad.
2. RUFO-AXILLARIS. pull.

The Screaming Cow-bird is larger than the allied species. The female is less than the male in size, but in colour they are alike, the entire plumage being deep blue-black, glossy, and with purple reflections; and under the wing at the joint there is a small rufous spot. The beak is very stout, the plumage loose, and with a strong, musky smell; the œsophagus remarkably wide.

It is far less common than the other species of *Molothrus*, but is not rare, and ranges south to the Buenos-Ayrean pampas, where a few individuals are usually found in every large plantation; and, like the *M. badius*, it remains with us the whole year. It is not strictly gregarious, but in winter goes in parties, never exceeding five or six individuals, and in the breeding-season in pairs. One of its most noteworthy traits is an exaggerated hurry and bustle thrown into all its movements. When passing from one branch to another, it goes by a series of violent jerks, smiting its wings loudly together; and when a party of them return from the fields they rush wildly and loudly screaming to the trees, as if pursued by a bird of prey. They are not singing-birds; but the male sometimes, though rarely, attempts a song, and utters, with considerable effort, a series of chattering unmelodious notes. The chirp with which he invites his mate to fly has the sound of a loud and smartly-given kiss. His warning or alarm-note when approached in the breeding-season has a soft and pleasing sound; it is, curiously enough, his only mellow expression. But his most common and remarkable vocal performance is a cry beginning with a hollow-sounding internal note, and swelling into a sharp metallic ring; this is uttered with tail and wings spread and depressed, the whole plumage raised like that of a strutting turkey-cock, whilst the bird hops briskly up and down on its perch as if dancing. From its puffed-out appearance, and from the peculiar character of the sound it emits, I believe that, like the Pigeon and some other species, it has the faculty of filling its crop with air, to use it as a "chamber of resonance." The note I have described is quickly and invariably followed by a scream, harsh and impetuous, uttered by the female, though both notes always sound as if proceeding from one bird. When on the wing the birds all scream together in concert.

The food of this species is chiefly minute seeds and tender buds; they also swallow large caterpillars and spiders, but do not, like their congeners, eat hard insects.

I became familiar, even as a small boy, with the habits of the Screaming Cow-bird, and before this species was known to naturalists, but could never find its nest, though I sought diligently for it. I could never see the birds collecting materials for a nest, or feeding their grown-up young like other species, and this might have made me suspect that they did not hatch their own eggs; but it never occurred to me that the bird was parasitical, I suppose because in summer they are always seen in pairs, the male and female being inseparable. Probably this is the only parasitical species in which there is conjugal fidelity. I also noticed that, when approached in the breeding-season, the pair always displayed great excitement and anxiety, like birds that have a nest, or that have selected a site on which to build one. But year after year the end of the summer would arrive, the birds reunite in parties of half a dozen, and the mystery remain unsolved. At length, after many years, fortune favoured me, and while observing the habits of another species (*Molothrus badius*), I discovered by chance the procreant habits of the Screaming Cow-birds; and as these observations throw some light on the habits of *M. badius*, I think it best to transcribe my notes here in full.

A pair of Leñateros (*Anumbius acuticaudatus*) have been nearly all the winter building a nest on an acacia tree sixty yards from the house; it is about 27 inches deep, and 16 or 18 in circumference, and appears now nearly finished. I am sure that this nest will be attacked before long, and I have resolved to watch it closely.

September 28.—To-day I saw a Bay-wing (*M. badius*) on the nest; it climbed over it, deliberately inspecting every part with the critical air of a proprietor who had ordered its construction, taking up and re-arranging some sticks and throwing others away from the nest. While thus engaged, two common Cow-birds (*M. bonariensis*), male and female, came to the tree; the female dropped on to the nest, and began also to examine it, peering curiously into the entrance and quarrelling with the first bird. After a few minutes she flew away, followed by her glossy consort. The Bay-wing continued its strange futile work until the owners of the nest appeared, whereupon it hopped aside in its usual slow leisurely manner, sang for a few moments, then flew away. The similarity in the behaviour of the two birds struck me very forcibly; in the great interest they take in the nests of other birds, especially in large covered nests, the two species are identical. But when the breeding-season comes their habits begin to diverge: then the Common Cow-bird lays in nests of other species, abandoning its eggs to their care; while the Bay-wings usually seize on the nests of other birds and rear their own young. Yet, as they do occasionally build a neat elaborate nest for themselves, the habit of taking possession of the nests of other birds is, most likely, a recently acquired one, and probably its tendency is to eradicate the original building instinct.

October 8.—This morning, while reading under a tree, my attention was aroused by a shrill note, as of a bird in distress, issuing from the neighbourhood of the Leñatero's nest; after hearing it repeated at intervals for over twenty minutes, I went to ascertain the cause. Two Bay-wings flew up from the ground under the nest, and on searching in the rank clover growing under the tree, I discovered the female Leñatero, with plumage wet and draggled, trembling and appearing half dead with the rough treatment she had experienced. I put her in the sun, and after half an hour, hearing her mate calling, she managed to flutter feebly away to join him. The persecutors had dragged her out of the nest, and would, no doubt, have killed her, had I not come so opportunely to the rescue.

Since writing the above, I have continued to watch the nest. Both the Bay-wings and Leñateros left it for some days. Six days after picking up the ill-treated female, the Leñateros came back and resumed possession. Four days later the Bay-wings also came back; but on finding the nest still occupied, they took possession of an unfinished oven of an Oven-bird on another tree within twenty yards of the first, and immediately began carrying in materials with which to line it. When they had finished laying I took their five eggs, at the same time throwing down the oven, and waited to see what their next move would be. They remained on the spot singing incessantly, and still manifesting anxiety when approached. I observed them four days, and then was absent from home as many more; on returning, I found that the Leñateros had once more disappeared, and that the nest was now held by the Bay-wings. I also noticed that they had opened an entrance very low down at the side of the nest which they were using; no doubt they had killed and thrown out the young Leñateros.

It was now early in November, the height of the breeding-season, and numbers of Common Cow-birds constantly visited the nest; but I was particularly interested in a pair of Screaming Cow-birds that had also began to grow fond of it, and I resolved to watch them closely. As they spent so much of their time near the nest, showing great solicitude when I approached it, I strongly hoped to see them breed in it, if the Bay-wings could only be got rid of. The Screaming Cow-birds would not, or dared not, attack them; and, as I always think that the worst possible use one can put a little bird to is to shoot it, I could not help them by destroying the Bay-wings. I therefore resolved to take their eggs, hoping that that would cause them to leave in disgust.

When I was satisfied from their movements that they had finished laying, I got up to the nest, and was astonished to find *ten* eggs instead of five, as I had confidently expected; for, though the Common Cow-birds had paid a great deal of attention to the nest, I knew the Bay-wings would not allow *them* to lay in it.

The ten eggs in the nest were all unmistakably Bay-wing's eggs; and having observed before that several females do occasionally lay together, I concluded that in this case two females had laid in the nest, though I had only seen two birds—male and female. After taking the ten eggs the Bay-wings still remained, and in a very short time they appeared to be laying again. When I had reason to think that the full complement was laid, I visited the nest and found five eggs in it; these I also took, and concluded that the second female had probably gone away, after having been deprived of her first clutch. During all this time the Screaming Cow-birds remained in the neighbourhood and occasionally visited the tree; but to my very great surprise the Bay-wings still stubbornly remained, and by-and-by I found that they were going to lay again—the fourth time! When I next visited the nest there were two eggs in it; I left them and returned three days later, expecting to find five eggs, but found seven! certainly more than one female had laid in the nest on this occasion. After taking these last seven eggs the Bay-wings left; and though the Screaming Cow-birds continued to make occasional visits to the nest, to my great disappointment they did not lay in it.

April 12.—To-day I have made a discovery, and am as pleased with it as if I had found a new planet in the sky. The mystery of the Bay-wings' nest twice found containing over the usual complement of eggs is cleared up, and I have now suddenly become acquainted with the procreant instinct of the Screaming Cow-bird. I look on this as a great piece of good fortune; for I had thought that the season for making any such discovery was already over, as we are so near to winter.

The Bay-wings are so social in their habits that they always appear reluctant to break up their companies in the breeding-season; no sooner is this over, and while the young birds are still fed by the parents, all the families about a plantation unite into one flock. About a month ago all the birds about my home had associated in this way together, and went in a scattered flock, frequenting one favourite feeding-spot very much, a meadow about fifteen minutes' walk from the house. The flock was composed, I believe, of three families, sixteen or eighteen birds in all: the young birds are indistinguishable from the adults; but I knew that most of these birds were young hatched late in the season, from their incessant strident hunger-notes. I first observed them about the middle of March. A week ago, while riding past the meadow where they were feeding, I noticed among them three individuals with purple spots on their plumage. They were at a distance from me, and I naturally concluded that they were young Common Cow-birds (*M. bonariensis*), casually associating with the Bay-wings. I was surprised to see them, for the young male *M. bonariensis* always acquires the purple plumage before March, so that these individuals were changing colour five weeks after the usual time. To-day, while out with my gun, I came upon the flock, and noticed four of the birds assuming the purple plumage, two of them being almost entirely that colour; but I also noticed with astonishment that they had bay- or chestnut-coloured wings, also that those with least purple on them were marvellously like the Bay-wings in the mouse-coloured plumage of the body and the dark tail. I had seen these birds before the purple plumage was acquired, and there was then not the slightest difference amongst them, the adults and their supposed offspring being alike; now some of them appeared to be undergoing the process of a transmutation into another species! I at once shot the four spotted birds along with two genuine Bay-wings, and was delighted to find that the first were young Screaming Cow-birds.

I must now believe that the extra eggs twice found in the nest of the Bay-wings were those of the Screaming Cow-bird, that the latter species lays chiefly in the nests of the former, that the eggs of the two species are identical in form, size, and colour, each bird also laying five, and that, stranger still, the similarity is as perfect in the young birds as it is in the eggs.

April 15.—This morning I started in quest of the Bay-wings, and observed one individual, that had somehow escaped detection the day before, assuming the purple dress. This bird I shot; and after the flock had resettled a short distance off, I crept close up to them, under the shelter of a hedge, to observe them more narrowly. One of the adults was closely attended by three young birds; and these all, while I watched them, fluttered their wings and clamoured for food every time the old bird stirred on its perch. The three young birds seemed precisely alike; but presently I noticed that one of them had a few minute purple spots, and on shooting this one I found it to be a young *M. rufoaxillaris*, while the other two were true young Bay-wings.

The hunger-cry of the young *M. badius* (Bay-wing) is quite different from that of the young *M. bonariensis*: the cry of the latter is a long, shrill, two-syllabled note, the last syllable being prolonged into a continuous squeal when the foster-parent approaches with food; the cry of the young *M. badius* is short, reedy, tremulous, and uninflected. The resemblance of the young *M. rufoaxillaris* to its foster-brothers in language and plumage is the more remarkable when we reflect that the adult bird in its habits, gestures, guttural notes, also in its deep purple plumage, comes much nearer to *M. bonariensis* than to *M. badius*. It seems impossible for mimicry to go further than this. A slight difference in size is quite imperceptible when the birds are flying about; while in language and plumage the keenest ornithologist would not be able to detect a difference. But it may be questioned whether this is really a case of an external resemblance of one species to another acquired by natural selection for its better preservation. Possibly the young *M. rufoaxillaris*, in the first stage of its plumage, exhibits the ancestral type—that of the progenitor of both species. If *M. badius* belonged to some other group—*Sturnella* or *Pseudoleistes*, for instance—it would scarcely be possible to doubt that the resemblance of the young *M. rufoaxillaris* to its foster-brothers resulted from mimicry; but as both species belong to the limited, well-defined group *Molothrus*, the resemblance may be ascribed to community of descent.

Formerly I believed that though *M. badius* is constantly seen rearing its own young, they also occasionally dropped their eggs in the nests of other birds. I could not doubt that this was the case after having witnessed a couple of their young following a Yellowbreast and being fed by it. I must now alter my opinion, for what then appeared to be proof positive is now no proof at all, for those two birds were probably the young of *M. rufoaxillaris*. There are, however, good reasons for believing that *M. rufoaxillaris* is parasitical almost exclusively on *M. badius*. I have spoken of the many varieties of eggs *M. bonariensis* lays. Those of *M. badius* are a trifle less in size, in form elliptical, densely and uniformly marked with small spots and blotches of dark reddish colour, varying to dusky brown; the ground-colour is white, but sometimes, though rarely, pale blue. It is not possible to confound the eggs of the two species. Now, ever since I saw, many years ago, the Yellowbreast feeding the supposed young Bay-wings, I have looked out for the eggs of the latter in other birds' nests. I have found hundreds of nests containing eggs of *M. bonariensis*, but never one with an egg of *M. badius*, and, I may now add, never one with an egg of *M. rufoaxillaris*. It is wonderful that *M. rufoaxillaris* should lay only in the nests of *M. badius*; but the most mysterious thing is that *M. bonariensis*, indiscriminately parasitical on a host of species, never, to my knowledge, drops an egg in the nest of *M. badius*, unless it be in a forsaken nest! Perhaps it will be difficult for naturalists to believe this; for if the *M. badius* is so excessively vigilant and jealous of other birds approaching its nest as to succeed in keeping out the subtle, silent, grey-plumaged, omnipresent female *M. bonariensis*, why does it not also keep off the far rarer, noisy, bustling, conspicuously coloured *M. rufoaxillaris*? I cannot say. The only explanation that has occurred to me is that *M. badius* is sagacious enough to distinguish the eggs of the common parasite, and throws them out of its nest. But this is scarcely probable, for I have hunted in vain under the trees for the ejected eggs; and I have never found the eggs of *M. badius* with holes pecked in the shells, which would have been the case had a *M. bonariensis* intruded into the nest.

With the results just recorded I felt more than satisfied, though so much still remained to be known; and I looked forward to the next summer to work out the rich mine on which I had stumbled by chance. Unhappily, when spring came round again ill-health kept me a prisoner in the city, and finding no improvement in my condition, I eventually left Buenos Ayres at the close of the warm season to try whether change of climate would benefit me. Before leaving, however, I spent a few days at home, and saw enough then to satisfy me that my conclusions were correct. Most of the birds had finished breeding, but while examining some nests of *Anumbius* I found one which Bay-wings had tenanted, and which for some reason they had forsaken leaving *ten* unincubated eggs. They were all like Bay-wings' eggs, but I have no doubt that five of them were eggs of *M. rufoaxillaris*. During my rides in the neighbourhood I also found two flocks of Bay-wings, each composed of several families, and amongst the young birds I noticed several individuals beginning to assume the purple plumage, like those of the previous autumn. I did not think it necessary to shoot more specimens.

The question, why *M. badius* permits *M. rufoaxillaris* to use its nest, while excluding the allied parasite, *M. bonariensis*, must be answered by future observers; but before passing from this very interesting group (*Molothrus*) I wish to make some general remarks on their habits and their anomalous relations to other species.

It is with a considerable degree of repugnance that we regard the parasitical instincts in birds; the reason it excites such a feeling is manifestly because it presents itself to the mind as—to use the words of a naturalist of the last century, who was also a theologian, and believed the Cuckoo had been created with such a habit—"a monstrous outrage on the maternal affection, one of the first great dictates of nature." An *outrage*, since each creature has been endowed with this all-powerful affection for the preservation of its own, and not another, species; and here we see it, by a subtle process, an unconscious iniquity, turned from its purpose, perverted and made subservient to the very opposing agency against which it was intended as a safeguard! The formation of such an instinct seems indeed like an unforeseen contingency in the system of nature, a malady strengthened, if not induced, by the very laws established for the preservation of health, and which the *vis medicatrix* of nature is incapable of eliminating. Again, the egg of a parasitical species is generally so much larger, differing also in coloration from the eggs it is placed with, whilst there is such an unvarying dissimilarity between the young bird and its living or murdered foster-brothers, that, unreasoning as we know instinct, and especially the maternal instinct, is, we are shocked at so glaring and flagrant an instance of its blind stupidity.

In the competition for place, the struggle for its existence, said with reason to be most deadly between such species as are most nearly allied, the operations are imperceptible, and the changes are so gradual, that the diminution and final disappearance of one species is never attributed to a corresponding increase in another more favoured species over the same region. It is not as if the regnant species had invaded and seized on the province of another, but appears rather as if they had quietly entered on the possession of an inheritance that was theirs by right. Mighty as are the results worked out by such a process, it is only by a somewhat strained metaphor that it can be called a *struggle*. But even when the war is open and declared, as between a raptorial species and its victims, the former is manifestly driven by necessity. And in this case the species preyed on are endowed with peculiar sagacity to escape its persecutions; so that the war is not one of extermination, but, as in a border war, the invader is satisfied with carrying off the weak and unwary stragglers. Thus the open, declared enmity is in reality beneficial to a species; for it is sure to cut off all such individuals as might cause its degeneration. But we can conceive no necessity for such a fatal instinct as that of the Cuckoo and Cow-bird destructive to such myriads of lives in their beginning. And inasmuch as their preservation is inimical to the species on which they are parasitical, there must also here be a struggle. But what kind of struggle? Not as in other species, where one perishes in the combat that gives greater strength to the victor, but an anomalous struggle in which one of the combatants has made his adversary turn his weapons against himself, and so seems to have an infinite advantage. It is impossible for him to suffer defeat; and yet, to follow out the metaphor, he has so wormed about and interlaced himself with his opponent that as soon as he succeeds in overcoming him he also must inevitably perish. Such a result is perhaps impossible, as there are so many causes operating to check the undue increase of any one species: consequently the struggle, unequal as it appears, must continue for ever. Thus, in whatever way we view the parasitical habit, it appears cruel, treacherous, and vicious in the highest degree. But should we attempt to mentally create a perfect parasitical instinct (that is, one that would be thoroughly efficient with the least possible prejudice to or injustice towards another species; for the preservation of the species on which the parasite is dependent is necessary to its own) by combining in imagination all known parasitical habits, eliminating every offensive quality or circumstance, and attributing such others in their place as we should think fit, our conception would probably still fall short in simplicity, beauty, and completeness of the actual instinct of *M. rufoaxillaris*. Instead of laying its eggs promiscuously in every receptacle that offers, it selects the nest of a single species; so that its selective instinct is related to the adaptive resemblance in its eggs and young to those of the species on which it is parasitical. Such an adaptive resemblance could not of course exist if it laid its eggs in the nests of more than one species, and it is certainly a circumstance eminently favourable to preservation. Then, there not being any such incongruity and unfitness as we find in nests into which other parasites intrude, there is no reason here to regard the foster-parents' affection as blind and stupid; the similarity being close enough to baffle the keenest sagacity. Nor can the instinct here appear in the light of an outrage on the maternal affection; for the young *M. rufoaxillaris* possesses no advantage over its foster-brothers. It is not endowed with greater strength and voracity to monopolize the attentions of the foster-parent or to eject the real offspring; but being in every particular precisely like them, it has only an equal chance of being preserved. To this wonderful parasitical instinct we may well apply Darwin's words, when speaking of the architecture of the hive-bee:—"Beyond this stage of perfection natural selection could not lead."

---

96. MOLOTHRUS BADIUS, Vieill.

(BAY-WINGED COW-BIRD.)

[PLATE VI. FIG. 1.]

Molothrus badius, *Burm. La-Plata Reise*, ii. p. 495 (Paraná and Tucuman). *Scl. et Salv. Nomencl.* p. 37; *Hudson, P. Z. S.* 1874, p. 163 (Buenos Ayres); *Durnford, Ibis*, 1877, p. 174 (Buenos Ayres); *Scl. Cat. B.* xi. p. 338.

Description.—Dull grey, beneath rather paler; wings chestnut; tips of primaries, inner portions of secondaries, and tail blackish; bill and feet black: total length 7·6 inches, wing 3·5, tail 3·0. *Female* similar.

Hab. Argentina, Paraguay, and Bolivia.

In this species the sexes are alike; the plumage of the body is grey-drab colour, with a black spot between the eye and beak; tail dark, the quills cinnamon-colour; beak and legs black. Azara describing it under the name of *Tordo pardo roxiso*, says it is a rare bird, so that it has probably greatly increased since his time, as it is now quite common in the Plata district.

The Bay-wings usually go in small flocks, numbering from ten to thirty individuals, and are not migratory, but in winter they travel about a great deal from place to place without extending their journeys more than a few miles in any direction. They are fond of coming about houses, and are frequently seen pecking at the fresh meat hanging out of doors; and, like other birds of the same tribe, feed chiefly on the ground. They spend a great portion of their time on trees, are familiar with man and inactive, and in their motions singularly slow and deliberate. Their language is varied. Curiosity or alarm is expressed by trilling notes, and before quitting a tree all the birds of a flock ceremoniously invite each other to fly with long clear notes, powerful enough to be heard a quarter of a mile away.

They also sing a great deal in all seasons, the song being composed of soft, clear, rather sweet notes, variously modulated, uttered in a leisurely manner, and seeming to express a composed frame of mind, all the birds in a flock singing in concert. During the cold season the flock always finds some sheltered sunny spot on the north side of a wood-pile or hedge, where they spend several hours every day, sitting still and singing in their usual quiet, soft style.

Their extreme sociability affects their breeding-habits, for sometimes the flock does not break up in spring, and several females lay in one nest together; but whether the birds are paired or practice a promiscuous intercourse, I have not been able to discover. They have a great partiality for the large domed nests made by the *Anumbius acuticaudatus*, called Leñatero in the vernacular. One summer a flock of about ten Bay-wings took possession of a Leñatero's nest on one of my trees, and after a few days I took fourteen eggs from it. Though the birds hopped, chirping round me, manifesting great solicitude, the eggs were quite cold, and had I left them many more would have been laid, no doubt; but as they were piled up three or four deep in the nest they could never have been hatched.

As a rule, however, the flock breaks up into pairs; and then a neat, well-made nest is built in the fork of a branch, lined with horsehair; or, oftener still, a Leñatero's nest is seized, the Bay-wings fighting with great spirit to get possession, and in it, or on it, their own nest is made. Like their relations, the Common Cow-bird, they seem strongly attracted by domed nests, and yet shrink from laying in the dark interior; as a rule when they have captured a Leñatero's nest they break a hole in the side and so admit the light and form an easy entrance. One summer a pair of Bay-wings attacked a Leñatero's nest on one of my trees; the fighting was kept up for three or four days, and then at the foot of the tree I found five young Leñateros, fully fledged, which had been pecked to death and thrown out of the nest.

The eggs of the Bay-wing are five in number, nearly round, and densely marked with dusky reddish brown.

Once I observed two young Bay-wings following a Yellow-breast (*Pseudoleistes virescens*) with their usual peculiar hunger-cry, and while I watched them they were fed several times by their foster-parents. Naturally I concluded that the Bay-winged Cow-bird is sometimes parasitical on other species, but I never saw anything afterwards to confirm me in that belief, and I believe now that I was mistaken, and that the young Bay-wings were not *real* Bay-wings, but the young of *Molothrus rufoaxillaris*.

---

97. AGELÆUS THILIUS (Mol.).

(YELLOW-SHOULDERED MARSH-BIRD.)

Agelæus thilius, *Burm. La-Plata Reise*, ii. p. 492 (Mendoza, S. Juan, Catamarca); *Scl. et Salv. Nomencl.* p. 37; *Durnford, Ibis*, 1876, p. 159 (Buenos Ayres), 1877, p. 33 (Chupat), p. 174 (Buenos Ayres), et 1878, p. 394 (Chupat); *White, P. Z. S.* 1882, p. 602 (Buenos Ayres); *Döring, Exp. al Rio Negro, Zool.* p. 40 (Colorado); *Barrows, Bull. Nutt. Orn. Cl.* viii. p. 134 (Entrerios); *Scl. Cat. B.* xi. p. 343.

Description.—Black; lesser upper and under wing-coverts yellow; bill and feet black: whole length 5·5 inches, wing 3·6, tail 2·7. *Female*: above pale brown striated with black; distinct superciliaries white; beneath paler, cinereous white with black striations; smaller, and bill shorter.

Hab. S. Peru, Chili, Paraguay, and Argentina.

This bird is abundant everywhere on the pampas, and does not migrate, but inhabits marshy situations in summer, building its nest amongst the rushes, and in winter ranges over the country. The male is entirely of an intense black, except the shoulders, which are pure yellow; the female is dull grey with fuscous markings, and, as was long ago remarked by Azara, the grey-plumaged are very much more numerous than the black individuals. The young birds are like the females, and possibly do not acquire the full black plumage until the second year, which would account for the great number of grey birds.

These birds are extremely sociable, being seen in flocks all the year round, even during the breeding-season; in winter a great many males separate themselves from the females, and are found associating together in flocks of from thirty to forty individuals.

They feed on the ground, keeping to the moist borders of marshes during summer; they avoid woods, but occasionally alight on trees, where they all sing in concert. The song, when an individual is heard singing alone, is, though limited in its range, very sweet, some of the notes being remarkable for their purity and expression. The bird sits on a rush or stalk while singing, and makes a long pause after every note or two, as if to make the most of its limited repertory. There is in the song one rich full note, which, to my mind, is unequalled for plaintive sweetness, and I am therefore surprised that Azara says only of this species that it sings passably well—'*canta razonablemente*.'

The nest is neatly made of dry grasses, and attached to the rushes growing in the water. The eggs are four, pointed, and spotted at the larger end with dull brown and black on a white ground.

---

98. AGELÆUS FLAVUS (Gm.).

(YELLOW-HEADED MARSH-BIRD.)

Xanthosomus flavus, *Scl. et Salv. P. Z. S.* 1869, p. 632 (Buenos Ayres); *iid. Nomencl.* p. 37; *Durnford, Ibis*, 1878, p. 59 (Buenos Ayres); *Döring, Exp. al Rio Negro, Zool.* p. 41 (Carhué); *Barrows, Bull. Nutt. Orn. Cl.* viii. p. 134 (Entrerios). **Agelæus flavus**, *Scl. Cat. B.* xi. p. 346.

Description.—Black; head and rump, bend of the wing, and body beneath bright yellow; bill and feet black: whole length 6·7 inches, wing 4·2, tail 3·2. *Female*: above brown, slightly striated; eyebrows, rump, and body beneath yellowish; bill and feet brown.

Hab. Paraguay, Uruguay, and Argentina.

Azara called this bird *Cabeza amarilla*, or Yellow-head. It is found throughout the eastern provinces of the Argentine country, ranging south to about the thirty-sixth degree of latitude, and is also common in the Banda Oriental.

The dull-plumaged birds are always very much more numerous than the bright-coloured males, though Azara strangely asserts that the sexes are alike. In Buenos Ayres, where it is called "Naranjo" by the country people in allusion to its orange tints, it is very well known on account of its yellow plumage, which looks so wonderfully brilliant in the sunshine, and its partiality for cultivated districts, where it follows the plough to pick up worms, and frequents the orchard to sing, associating with the common Cow-bird and Yellow-breast. It remains all the year, and is very sociable, going in flocks of from twenty to fifty individuals, which when they settle on the trees all sing in concert, pouring out their few peculiar notes with great power and emphasis.

Even in the breeding-season these companies do not always break up, and frequently several pairs have nests near together. The nest is usually built in a cardoon thistle, two or three feet above the ground, and is made of dry grass. The eggs are four, pointed, white or with a bluish tinge, and speckled irregularly with deep brown, the spots being closer and sometimes confluent at the broad end.

Concerning the plumage of this species Mr. Barrows writes:—"Late in March, 1881, we found this species in large flocks on the Pigué, and it was a beautiful sight to see a hundred or more fluttering about among the snowy plumes of the pampas grass, and displaying their rich black and yellow dress. Unlike most other birds obtained at that time, their plumage seemed nearly as bright and fresh as in summer."

---

99. AGELÆUS RUFICAPILLUS, Vieill.

(RED-HEADED MARSH-BIRD.)

Agelæus ruficapillus, *Scl. Cat. B.* xi. p. 347. **Xanthosomus ruficapillus,** *Scl. et Salv. P. Z. S.* 1869, p. 159 (Buenos Ayres); *iid. Nomencl.* p. 37; *White, P. Z. S.* 1882, p. 602 (Salta, Catamarca). **Chrysomus frontalis,** *Burm. La-Plata Reise,* ii. p. 492 (Paraná).

Description.—Glossy blue-black; crown of head and middle of throat dark chestnut; bill and legs black: whole length 7·5 inches, wing 3·7, tail 2·9.

Hab. Argentina and Paraguay.

The sexes are alike in this species: the crown of the head is rufous, and with this exception the whole plumage a rich glossy blue-black. The beauty of the bird and its delicate, plaintive voice would no doubt make it a favourite with man if he saw more of it, only it lives and breeds in marshes, and does not come near his habitations. The Red-heads are gregarious and migratory. The flock can scarcely be said to break up in the breeding-season, as the birds all make their nests near together in the reeds. The nest is placed about one or two feet above the water, is about six inches in depth, and made of leaves and aquatic grasses woven together. The eggs are four, pointed, with a white or pale bluish ground, and spotted with black at the larger end.

The song of the Red-head is quite unique in character. It begins with a low, hollow-sounding note, then the voice changes to a clear, sorrowful tone, rising in a rapid succession of short notes, and falling again in longer ones.

After the breeding-season the birds fly about in flocks of two or three hundred individuals, and sing in concert on the trees.

Their chirp has a peculiar metallic sound, and can be imitated by tapping on the edge of a copper bell with the finger-nail.

---

100. LEISTES SUPERCILIARIS, Bp.
(RED-BREASTED MARSH-BIRD.)

Leistes superciliaris, *Hudson, P. Z. S.* 1870, p. 333 (Buenos Ayres); *Scl. et Salv. Nomencl.* p. 38; *Durnford, Ibis,* 1877, p. 175 (Buenos Ayres); *White, P. Z. S.* 1882, p. 602 (Salta); *Barrows, Bull. Nutt. Orn. Cl.* viii. p. 136 (Entrerios); *Scl. Cat. B.* xi. p. 349. **Trupialis guianensis,** *Burm. La-Plata Reise,* ii. p. 490 (Banda Oriental).

Description.—Brownish black; superciliaries pale brown; bend of the wing and body beneath from the chin to the middle of the belly bright scarlet; bill and legs black: length 7·0 inches, wing 4·0, tail 2·5. *Female* pale brown, above variegated with black, below stained with red on the breast; flanks and lower belly striated with black; tail cinereous brown, with black cross bars.

Hab. Argentina, Paraguay, and E. Brazil.

The most interesting point concerning this species is the very great difference in habits, as well as appearance, existing between the sexes. In form it resembles the Starling of Europe, but is a trifle smaller and has a shorter tail. The male is black, the upper parts faintly mottled with yellowish grey; there is a straw-coloured stripe over the eye; the throat and breast bright crimson. The female is a smaller bird, and in colour dull fulvous grey, mottled with fuscous; the red tint on the breast scarcely perceptible.

These birds are migratory, and appear everywhere in the eastern part of the Argentine country early in October, arriving singly, after which each male takes up a position in a field or open space abounding with coarse grass and herbage, where he spends most of the time perched on the summit of a tall stalk or weed, his glowing crimson bosom showing at a distance like some splendid flower above the herbage. At intervals of two or three minutes he soars vertically up to a height of twenty or twenty-five yards to utter his song, composed of a single long, powerful, and rather musical note, ending with an attempt at a flourish, during which the bird flutters and turns about in the air; then, as if discouraged at his failure, he drops down, emitting harsh guttural chirps, to resume his stand. Meanwhile, the female is invisible, keeping closely concealed under the long grass. But at length, attracted perhaps by the bright bosom and aerial music of the male, she occasionally exhibits herself for a few moments, starting up with a wild, zigzag flight, like a Snipe flushed from its marsh, and, darting this way and that, presently drops into the grass once more. The moment she appears above the grass the male gives chase, and they vanish from sight together. Thus, while in colour, habits, language, and even in its manner of soaring up like a rocket to let off its curious melody, the male is the most conspicuous of small birds, the female, acted on in an opposite direction by natural selection, has been, so to speak, effaced. While flying, they do not look like birds of the same species: the male moves with wings rapidly fluttered, like a Starling, but with a slower, more laborious flight, and without deviating; the female, in her eccentric movements in the air, reminds one of a large moth driven from its hiding-place, and flying about confused with the glare of noon.

The nest is made of dry grass on the ground, so cunningly concealed that it is most difficult to find. The eggs are four, white, spotted with reddish brown. When they have young, I have never been able to detect the female flying about in search of food.

All through the summer these birds are solitary, but when migrating in the autumn, though many are seen travelling singly, and appear very conspicuous as they fly laboriously in a straight line, at an altitude of about twenty yards from the surface, others are seen making their journey in small flocks or parties composed of six to a dozen individuals. These are the males. The females travel separately, in twos or threes or singly, flying nearer to the earth, with frequent pauses, when the wings cease beating, and intervals of gliding, also darting occasionally to one side, as if the bird had suddenly taken fright.

---

101. AMBLYRHAMPHUS HOLOSERICEUS (Scop.).

(SCARLET-HEADED MARSH-BIRD.)

Amblyrhamphus holosericeus, *Hudson, P. Z. S.* 1870, p. 114 (Buenos Ayres); *Scl. et Salv. Nomencl.* p. 37; *Durnford, Ibis,* 1877, p. 174 (Buenos Ayres); *Gibson, Ibis,* 1880, p. 18 (Buenos Ayres); *White, P. Z. S.* 1882, p. 602 (Buenos Ayres); *Barrows, Bull. Nutt. Orn. Cl.* viii. p. 135 (Entrerios); *Scl. Cat. B.* xi. p. 351. **Amblyrhamphus ruber,** *Burm. La-Plata Reise,* ii. p. 491 (Entrerios, Santa Fé, Paraná).

Description.—Black; whole head and neck all round, and upper breast and thighs scarlet; bill and feet black: total length 9·5 inches, wing 4·5, tail 4·0. *Female* similar. *Young* uniform black.

Hab. Paraguay, Uruguay, and Argentina.

Azara named this species *Tordo negro cabeza roxa;* it is also called *Boyero* (ox-herd) by country people, from its note resembling the long whistle of a drover; and sometimes *Chisel-bill,* from the peculiar conformation of the beak, which is long, straight, and has a broad fine point like a chisel. In both sexes the plumage of the head and neck is scarlet, of an exceedingly brilliant tint, all other parts intense black.

These birds are lively, active, and sociable, going in flocks of from half-a-dozen to thirty individuals; they remain all the year, and inhabit the marshes, from which they seldom wander very far, but seek their insect food in the soft decaying rushes. They are common on the swampy shores of the Plata, and when seen at a distance, perched in their usual manner on the summits of the tall rushes, their flame-coloured heads shine with a strange glory above the sere sombre vegetation of the marshes. The long whistling note above mentioned is their only song, but it varies considerably, and often sounds as mellow and sweet as the whistle of the European Blackbird.

The nest is an ingenious structure of dry grasses, fastened to the upright stems of an aquatic plant, three or four feet above the water. The eggs are four, in size and form like those of the English Song-Thrush, spotted somewhat sparsely with black on a light blue ground.

The young birds are entirely black at first, and afterwards assume on the head and neck a pale terra-cotta red, which gradually deepens to vivid scarlet.

---

102. PSEUDOLEISTES VIRESCENS (Vieill.).

(YELLOW-BREASTED MARSH-BIRD.)

Pseudoleistes virescens, *Hudson, P. Z. S.* 1870, p. 549, et 1874, p. 156 (Buenos Ayres); *Scl. et Salv. Nomencl.* p. 37; *Durnford, Ibis,* 1877, p. 175, et 1878, p. 59 (Buenos Ayres); *Gibson, Ibis,* 1880, p. 31 (Buenos Ayres); *White, P. Z. S.* 1882, p. 602 (Buenos Ayres); *Barrows, Bull. Nutt. Orn. Cl.* viii. p. 135 (Entrerios); *Scl. Cat. B.* xi. p. 352. **Leistes anticus**, *Burm. La-Plata Reise,* ii. p. 491 (Paraná).

Description.—Above and below dark olive-brown; lesser upper wing-coverts, under wing-coverts, and middle of the abdomen yellow; bill black; feet dark brown: total length 9·5 inches, wing 4·6, tail 3·8. *Female* similar.

Hab. S. Brazil, Paraguay, Uruguay, and Argentina.

In both sexes in this species the plumage is deep olivaceous brown, the breast pure yellow. It is active, strong on the wing, sociable and noisy; and being moreover a pretty and elegant bird, very common in settled districts, and with a preference for man's neighbourhood, it is familiar to every one, and has won amongst many competitors the vernacular name of *Pecho-amarillo* (Yellow-breast), for with us yellow-breasted species are somewhat numerous. It remains all the year, invariably going about in flocks of from twenty to thirty birds, and feeds on the ground in the fields or on the open plain. While they are feeding, one bird takes up a position on a stalk or thistle-top to keep guard; when he flies down another bird takes his place; if a person approaches, the sentinel gives the alarm, and all the birds fly off in a very close flock, making the air resound with their loud ringing notes. After feeding, they repair to the trees, where they join their robust voices in a spirited concert, without any set form of melody such as other song-birds possess, but all together, flinging out their notes at random, as if mad with joy. In this delightful hubbub there are some soft silvery sounds. Where they are never persecuted they have little fear of man, but they invariably greet his approach with a loud vigorous remonstrance.

In October the birds break up their companies to pair. Sometimes they breed on the open plain in a large cardoon thistle, but a thick bush or low tree is preferred. The nest is like that of a Thrush, being deep, compactly made of dry grass and slender sticks, plastered inside with mud, and lined with hair or soft dry grass. It is, however, deeper and more symmetrical than the Thrush's nest, and it is sometimes plastered with cow-dung instead of with mud. The eggs are four, very long, white, and abundantly spotted with deep red, the spots becoming confluent at the large end.

The Yellow-breast is never seen to quarrel with its fellows or with other birds, and it is possibly due to its peaceful disposition that it is more victimized by the parasitical *Molothrus* than any other bird. I have frequently found their nests full of parasitical eggs, as many as fourteen and in one case sixteen, eggs in one nest. In some seasons all the nests I found and watched were eventually abandoned by the birds on account of the number of parasitical eggs dropped in them. I have also so frequently found parasitical eggs on the ground under the nest that I believe the Yellow-breast throws out some of these foreign eggs, and in one instance I was quite sure that this had happened. The nest was in a cardoon bush, and contained five eggs—two of the Yellow-breast and three parasitical. These three were of the variety most thickly mottled with red, and consequently closely resembling the eggs of the Yellow-breast. I was surprised to find five more eggs of the Cow-bird on the ground, close together, and about three feet from the bush; and these five eggs were all pure white and unspotted. Naturally I asked, How came these eggs in such a position? They had not fallen from the nest, which was very deep, contained few eggs, and was scarcely thirty inches above the ground. Then they were all white, while those in the nest were mottled. That the eggs had been laid in the nest I felt certain; and the only way I can account for their being in the place where I found them is that the Yellow-breast itself removed them, taking them up in its bill and flying with them to the ground. If I am right, we must believe that this individual Yellow-breast had developed an instinct unusual in the species, which enables it to distinguish, and cast out of its nest, eggs very different from its own—an instinct, in fact, the object of which would be to counteract the parasitical habit of *Molothrus*. What would be the effect of such an instinct should the species acquire it? Doubtless it would be highly prejudicial to the parasitical birds laying white eggs, but favourable to those laying mottled eggs. This would be natural selection operating in a very unusual manner; for the Yellow-breast, or other species, would improve another to its own detriment, since the more the parasitical eggs assimilated to its own, the greater would be the likelihood of their being preserved. The perfect similarity of the eggs of *M. rufoaxillaris* to those of *M. badius* perhaps was brought about in this way. But, it may be added, if besides the Yellow-breast some one other species laying very different eggs (a *Zonotrichia* or *Tyrannus*, for instance) should also acquire this distinguishing habit, and eject all eggs unlike its own from its nest, the habit in the two or more species would ultimately cause the extinction of the parasite.

It might throw some light on this obscure subject to examine, for several successive summers, a large number of nests, to ascertain whether the nests of the Yellow-breast are often found without any white unspotted eggs, or if the same proportional number of white (parasitical) eggs are found in the nests of the Yellow-breast, Scissor-tail, Song-Sparrow, Pipit, and other species.

---

103. TRUPIALIS MILITARIS (Linn.).

(PATAGONIAN MARSH-STARLING.)

Sturnella militaris, *Hudson, P. Z. S.* 1872, p. 548 (Rio Negro); *Scl. et Salv. Nomencl.* p. 38; *Durnford, Ibis,* 1877, p. 33, et 1878, p. 394 (Chupat). **Sturnella loica**, *Döring, Exp. al Rio Negro, Zool.* p. 41 (R. Colorado, R. Negro). **Trupialis militaris**, *Scl. Cat. B.* xi. p. 356. **Trupialis loyca**, *Burm. La-Plata Reise,* ii. p. 491 (Mendoza).

Description.—Above brown, especially on the head and back variegated with black; superciliaries in front of the eye red, behind the eye white; beneath black; throat, middle of the neck, breast, and upper belly scarlet; sides of belly and crissum with brownish edgings to the black feathers; under wing-coverts white; bend of the wing red; slight rictal stripe white; bill horn-colour; feet brown: total length 10·0 inches, wing 4·7, tail 3·6. *Female* similar.

Hab. Chili, Patagonia, and Falkland Islands.

Two species of *Trupialis* inhabit the southernmost part of the Argentine Republic, the present being confined to Patagonia and South

Chili, while its northern representative inhabits the pampas of Buenos Ayres and Uruguay. Probably the Colorado River, which separates two districts differing in soil and vegetation, is the boundary-line dividing their habitats. So nearly alike are these two birds in colour, language, and habits, that they seem rather like races than species; and they were so regarded by naturalists until recently, when the pampas bird was raised to the rank of a distinct species, with the name of *Trupialis defilippii*. Unfortunately the old name *militaris* fits the Pampas, and not the Patagonian, Starling best; but of this I shall speak when I describe the former species.

In form and size *T. militaris* resembles the Common Starling of Europe, but differs from it in habits, flight, language, and in colouring; its upper plumage being fuscous mottled with yellowish grey; the throat and bosom scarlet inclining to crimson. This hue varies greatly, the breast-feathers being often tipped with white, which subdues the intense red, and gives it a rosy tint in some individuals. The female is paler-plumaged than the male, and has less red on the breast.

It inhabits the whole of Patagonia to the Strait of Magellan, but is confined to the valleys or to the neighbourhood of water; and Durnford remarks that it is a useful bird to the traveller in that thirsty country, as its presence is a sure indication of water. It is resident, and is seen in small parties of four or five, or in small flocks seldom exceeding twenty or thirty in number. It feeds and lives on the ground, and only occasionally is it seen to perch on a low bush. Its flight is strong, and it flies about a great deal, and usually utters its song when on the wing. The song is continued all the year, and is heard even on the coldest days in winter; the notes are few and not highly melodious, but are cheerful and vigorous.

The nest is made of dry grass and rootlets attached to the rushes in moist ground, and placed close to or resting on the surface. The eggs are five, the ground-colour white spotted or blotched with reddish brown.

---

104. TRUPIALIS DEFILIPPII, Bp.

(DE FILIPPI'S MARSH-STARLING.)

Sturnella defilippii, *Scl. et Salv. P. Z. S.* 1869, p. 161 (Buenos Ayres); *iid. Nomencl.* p. 38; *Döring, Exp. al Rio Negro, Zool.* p. 41 (Pampas); *Barrows, Bull. Nutt. Orn. Cl.* viii. p. 136 (Pampas south of Buenos Ayres).
Trupialis defilippii, *Scl. Cat. B.* xi. p. 357. **Trupialis militaris**, *Burm. La-Plata Reise*, ii. p. 490 (Rosario, Paraná).

Description.—Like *Tr. militaris*, but rather smaller, and with the under wing-coverts black.

Hab. Northern Argentina and Uruguay.

Throughout the country where this species abounds it is called *Pecho colorado*, which is certainly better than Azara's barbarous, if picturesque, name of *Degollado*; but no happier name than *militaris* could have been invented for it, by which it was formerly known to naturalists; and though it was given to the bird merely on account of the red breast, and was therefore equally applicable to all the red-breasted species on the globe, in this case it accidentally seemed to describe a peculiar habit of the bird, as well as its bright livery.

In size, form, gait, flight, language, and colour the present bird very closely resembles the Patagonian Starling; but the crimson on the breast is brighter and the upper parts are darker. Its nesting-habits are also like those of the southern bird; the number and colour of the eggs being the same in both species. One trivial difference in habit is that De Filippi's Starling occasionally soars up a few yards into the air when uttering its song. In inhabits the moist grassy pampas in the southern part of the Buenos Ayrean province, and is there abundant and unites in large flocks. At the approach of the cold season there is a general movement northwards of the birds, which does not, however, extend far, as the birds, although strong fliers, travel slowly and in a peculiar manner; it is in this season, when the birds are seen moving in large flocks, that the name of Military Starling strikes one as being peculiarly appropriate. They do not journey through the air like other migrants, but move over the ground, when the flock, composed of four or five hundred to a thousand or more individuals, is extended so as to present a very long front, and at intervals the hindmost birds fly over the others and alight just in advance of them: the long front, the precision of their movements, and their scarlet bosoms all turned one way, suggest the idea of a disciplined army on its march.

They never perch on trees, but frequently alight on the roof of a rancho or other elevation affording a secure footing. They are tame birds and fly reluctantly; when approached, they usually crouch down, hiding their crimson bosoms, and remain motionless in order to escape observation. In disposition they are peaceful, and so fond of society that when one becomes separated from his fellows he will unite with birds of another kind, even with Plovers or Tyrant-birds.

On the great monotonous plains, where most of the small birds are grey- or brown-plumaged, and in winter when there are no flowers to satisfy the desire of the eye for bright colour, it is delightful, while travelling, to meet with an army of these Starlings: their crimson bosoms, less bright than the hues of some tropical species, seem then to glow with a strange splendour on the sombre green of earth, and the sight produces an exhilarating effect on the mind.

---

105. ICTERUS PYRRHOPTERUS, Vieill.
(CHESTNUT-SHOULDERED HANG-NEST.)

Icterus pyrrhopterus, *Scl. et Salv. P. Z. S.* 1868, p. 140; *iid. Nomencl.* p. 36; *Hudson, P. Z. S.* 1870, p. 88 (Buenos Ayres); *Durnford, Ibis,* 1878, p. 59; *Salvin, Ibis,* 1880, p. 356 (Salta); *Barrows, Bull. Nutt. Orn. Cl.* viii. p. 133 (Entrerios); *Scl. Cat. B.* xi. p. 368. **Xanthornus pyrrhopterus,** *Burm. La-Plata Reise,* ii. p. 493 (Paraná, Tucuman).

Description.—Uniform black; upper lesser wing-coverts chestnut; bill black; feet dark brown: whole length 7·7 inches, wing 3·6, tail 3·8. *Female* similar, but smaller.

Hab. S. Brazil, Paraguay, Bolivia, and Argentina.

This interesting bird, the only *Icterus* found in the Argentine Republic, ranges south to Buenos Ayres, where it is migratory, and appears in small flocks of six or eight individuals in September; but soon after arriving these little companies break up, and the birds are subsequently found singly or in pairs in the woods along the Plata River.

The sexes are alike in colour, but the male is considerably larger; the whole plumage is an intense black, excepting a rufous spot on the shoulder, seen only when the bird is on the wing; the bill is black and curved, the body slender, and the tail long. It is a loquacious bird, most of its tones being low and pleasing; exceedingly restless in disposition, incessantly passing from tree to tree, jerking its long tail and clinging to the branches in various attitudes, while searching for insects in the decayed bark. While thus engaged it utters a great variety of chirping and guttural sounds, interspersed with short agreeable notes. It also has a song of considerable merit, low and varied in tone, with a peculiar ventriloquism in many of the notes which produce a confusing idea on the listener that the bird approaches and recedes alternately whilst uttering them. While singing the bird continues moving, but always concealed in the thick foliage, and it is probably this constant turning about of the singer, and the notes coming through leafy screens of varying density, which makes the ventriloquism and gives so much light and shade to its mysterious melody.

The first bird of this species I shot was wounded very slightly in one wing and fell into a stream; to my very great surprise it began singing its usual song while floating about on the surface, making no attempt to swim. After fishing it out it continued to sing at intervals in my hand: how strange it was to hear this bleeding captive bird warbling out soft, sweet notes which seemed to express only pleasant emotions! Yet it was evident that the bird was fully alive to its danger, for it struggled violently to escape and bit my finger savagely with its sharp beak.

I subsequently found a nest; it was about seven inches deep, composed entirely of lichens gathered from the boles of trees, ingeniously woven together and suspended from the small twigs and leaves at the extremity of a branch. There were no eggs in it, but the birds fluttered in great trouble about me, and, what greatly surprised me, uttered a variety of *singing* notes, unlike their usual song, but many of them closely resembling the notes of other songsters, which made me think that this *Icterus* possesses the mimicking faculty to some extent. This, however, is a question it would be difficult to decide. It seems certain, however, that this species is incapable of expressing any distressing feeling, such as pain, fear, or parental anxiety, with loud harsh notes like other birds. It is much to be regretted that Azara, who found this species common in Paraguay, did not pay more attention to its habits and language, which make it specially interesting even in a family so rich in strange habits as the *Icteridæ*.

---

106. APHOBUS CHOPI (Vieill.).

(CHOPI BOAT-TAIL.)

Chopi, *Azara, Apunt.* i. p. 282. **Aphobus chopi**, *Scl. Cat. B.* xi. p. 405.
Psarocolius unicolor, *Burm. La-Plata Reise*, ii. p. 494 (Mendoza).

Description.—Uniform black, slightly resplendent; bill and feet black; lower mandible sulcated: whole length 9·2 inches, wing 5·2, tail 4·0. *Female* similar, but smaller and duller; bill not sulcated.

Hab. S. Brazil, Paraguay, Argentina, Bolivia, and Peru.

The Chopi, which is said to be quite common in Paraguay, is only found in the north-eastern part of the Argentine Republic, consequently I have never seen it, except as a cage-bird; nor is there anything about it in the notes of recent collectors and travellers who have visited the upper waters of the Plata. This is, however, not to be regretted, since Azara gave a very full and spirited account of this species in his 'Apuntamientos;' although it certainly does seem strange that the Chopi should have had two careful observers of its habits fully a century ago, namely, Azara and his friend and fellow-naturalist the priest Noseda, and not one since. Evidently Azara was very familiar with this bird, for he describes it lovingly and at great length, his history of it being one of the most charming things in his work. According to him the Chopi is a highly sagacious bird, and although a frequent visitor to courtyards and verandahs of houses in Paraguay, too shy and suspicious to be caught with snares. It has a strong and easy flight, and readily attacks any large bird passing near, following it persistently in the air, or, pouncing down, fastens itself on its enemy's back. If the Caracara Eagle (*Polyborus*) alights in order to shake off its persecutor, the Chopi perches at a distance of a few feet, where it assumes an indifferent manner; but no sooner does the Caracara allow its attention to wander from its adversary, then it is again subjected to fresh insult. These attacks on so large and powerful a species may be regarded as mere impertinences, but by practising them the Chopi is soon able to rid himself of the presence of any unwelcome bird. From a long distance he recognizes an enemy, by its figure, or even its shadow, and warns all birds of the coming danger with a loud whistle, which at once sends them into hiding, while the Chopi goes bravely out to the encounter; and the result is invariably a victorious song on his part, beginning with the sound of his own name, and running through a variety of whistled notes. He also sings well in captivity and when his mate is incubating; and his voice is first heard welcoming the dawn from the eaves and tiled roofs of houses where he roosts. The pairing-season is in November; and, Noseda adds, the breeding-place is a hole in a bank, or tree-trunks, or in a wall under the eaves, and occasionally the nest is made in the small branches of an orange or other close-leafed tree, and is built of sticks and straws carelessly disposed, with a few feathers for lining. The eggs are four, and white.

It may be added that between Azara and his friend Noseda there was a great controversy respecting the parasitical habits of the common Cow-bird (*Molothrus*), which were first discovered by the former, and disbelieved in by Noseda, who accounted for the fact that the Cow-bird is never seen to make a nest by supposing that species to be the year-old young of the Chopi, which, he further imagined, took three years to acquire the adult form and plumage. Such an idea might seem to discredit Noseda as a naturalist, if we did not remember that Gilbert White at the same period was trying to prove the hybernation of Swallows in England. The whole of the discussion appears in the 'Apuntamientos,' under the description of the Chopi; and Noseda is there allowed to state his own case; after which the better observer, Azara, gives five objections to the theory, any one of which would be sufficient to demolish it.

Fam. XII. CORVIDÆ, or CROWS.

The typical black Crows and Ravens do not penetrate into the New World so far south as South America, their range terminating in the highlands of Guatemala and the islands of the Greater Antilles. They have, however, as representatives in South America several allied groups, clad mostly in various shades of blue, and commonly known as Blue Crows and Blue Jays. Two of these are known to occur in the woods of the northern provinces of Argentina.

---

107. CYANOCORAX CHRYSOPS (Vieill.).

(URRACA JAY.)

Cyanocorax chrysops, *Sharpe, Cat. B.* iii. p. 120. **Cyanocorax pileatus,** *Burm. La-Plata Reise,* ii. p. 495 (Uruguay, Tucuman); *Scl. et Salv. Nomencl.* p. 39; *White, P. Z. S.* 1882, p. 602 (Salta); *Barrows, Bull. Nutt. Orn. Cl.* viii. p. 136 (Entrerios). **Cyanocorax tucumanus,** *Cab. J. f. O.* 1883, p. 216 (Tucuman).

Description.—Above black, slightly glossed with purple; head, throat, and breast black, the plumes of the crown forming a crest; nape greyish blue, deepening into purple on the hind neck; above the eye a blue spot; below the eye a second spot and a patch at the base of the lower mandible purple; under surface and tip of the tail creamy yellow; bill and feet black: total length 14 inches, wing 6·3, tail 7·1.

Hab. Brazil, Paraguay, Uruguay, and Northern Argentina.

"This Blue Jay is very common about the woods at Campo Colorado (Salta). It is remarkably tame and inquisitive, and has a great variety of peculiar grating metallic notes. On the sierras of Totoral I likewise met with it; and it is found as far south as Tucuman and Catamarca." (*White.*) According to Azara it is abundant in Paraguay and familiar with man, coming a great deal about the houses. The vernacular name there is *Acahé*.

---

108. CYANOCORAX CÆRULEUS (Vieill.).

(AZURE JAY.)

Cyanocorax cæruleus, *Sharpe, Cat. Birds,* iii. p. 126; *White, P. Z. S.* 1882, p. 603 (Misiones). **Cyanocorax azureus,** *Scl. et Salv. Nomencl.* p. 40.

Description.—Head all round, neck, and throat down to upper part of breast black; all the rest of the plumage fine and glossy blue; beak and feet black: total length 15·5 inches, wing 8·15, tail 6·9.

Hab. S. Brazil, Paraguay, and N. Argentina.

This fine species was originally discovered by Azara in Paraguay. In June 1881 specimens were obtained at San Javier, Misiones, by White. He writes: "This bird is found here at times in flocks, and makes a great deal of noise, but is exceedingly wild and difficult to shoot, so that I had to use ball-cartridge to secure this specimen."

Suborder II. *OLIGOMYODÆ.*

Fam. XIII. TYRANNIDÆ, or TYRANTS.

The Tyrannidæ or Tyrant-birds, which play the part of our Flycatchers in the New World, although they are quite different in structure, contribute an important element in the Neotropical Avifauna, numbering as they do more than 350 species. Of these 63 have already been met with within the limits of the Argentine Republic, and the list will, no doubt, be further extended, as many of these birds are small and dull in colour, and easily hide themselves in the dense forests to which they resort.

The pampas are the special home of the Tæniopterinæ, or "Walking Tyrants" as Swainson called them, which have strong feet and long tarsi, and frequent open spaces. Amongst them the genera *Tænioptera, Alectrurus, Cnipolegus,* and *Lichenops* may be regarded as specially characteristic of the Argentine avifauna. On the other hand, the Elaineinæ and Platyrhynchinæ, which are eminently birds of the dense forest, are much less numerous within the bounds of Argentina, and are mostly confined to the wooded districts of the north. The typical Tyranninæ form a more mixed group, some species of which, such as *Tyrannus melancholicus* and *Pyrocephalus rubineus,* are prominent objects in the Argentine ornis.

---

109. AGRIORNIS STRIATA, Gould.

(STRIPED TYRANT.)

Agriornis striatus, *Burm. La-Plata Reise,* ii. p. 459 (Cordova); *Salvin, Ibis,* 1880, p. 356 (Tucuman); *Döring, Exp. al Rio Negro, Zool.* p. 41 (R. Colorado, R. Negro).

Description.—Above cinereous; wings blackish cinereous with light edgings; tail dark cinereous, with a distinct white margin to the external rectrix, and slight whitish tips to all rectrices; lores blackish; short superciliaries white:

below pale cinereous; throat white, broadly striated with black; bill above black, below pale: whole length 9·0 inches, wing 4·5, tail 4·3. *Female* similar.

Hab. Eastern Argentina and Patagonia.

This species, originally discovered by Darwin in Patagonia, is also found in Eastern Argentina. White obtained it in Catamarca, Durnford in Tucuman, and Burmeister on the Rio Quarto, in Cordova.

---

110. AGRIORNIS MARITIMA (d'Orb. et Lafr.).

(WHITE-TAILED TYRANT.)

Agriornis maritima, *Scl. et Salv. Nomencl.* p. 41; *Durnford, Ibis,* 1878, p. 394 (Chupat); *Barrows, Bull. Nutt. Orn. Cl.* vol. viii. p. 137 (Sierra de la Ventana). **Agriornis leucurus,** *Burm. La-Plata Reise,* ii. p. 459 (Mendoza).

Description.—Above cinereous; secondaries edged externally and tipped with white; tail dark cinereous, the four external rectrices with the whole outer webs and about the one-third apical part of the inner webs white, the median rectrices tipped with white; below paler cinereous; throat greyish white, slightly striated with dark cinereous; lower belly, crissum, flanks, and under wing-coverts white, more or less tinged with cinnamomeous; bill and feet black: whole length 9·0 inches, wing 5·3, tail 4·0. *Female* similar.

Hab. Argentina, Patagonia, Chili, and Bolivia.

This *Agriornis* was obtained by Durnford at Tombo Point in Central Patagonia, by Burmeister in the Sierra de Uspallata, near Mendoza, and by Barrows in the rocky gorges of the Sierra de la Ventana.

Darwin tells us of this species that it "is a scarce, shy, solitary bird, frequenting the valleys in which thickets grow, but often feeding on the ground. In the interior plains of Patagonia, on the banks of the Santa Cruz, I several times saw it chasing beetles on the wing, in a peculiar manner, half hopping and half flying; when thus employed it spreads its tail, and the white feathers in it are displayed in a very conspicuous manner. I also met with the species in the lofty and arid valleys on the eastern side of the Cordillera of Central Chili, and likewise at Copiapo." (Zool. Voy. 'Beagle,' iii. p. 57.)

---

111. MYIOTHERETES RUFIVENTRIS (Vieill.).

(CHOCOLATE TYRANT.)

Myiotheretes rufiventris, *Scl. et Salv. P. Z. S.* 1868, pp. 138, 141 (Conchitas); *iid. Nomencl.* p. 42; *Durnford, Ibis,* 1877, p. 175 (Buenos Ayres), et 1878, p. 394 (Centr. Patagonia); *White, P. Z. S.* 1882, p. 603 (Buenos Ayres);

Barrows, Bull. Nutt. Orn. Cl. viii. p. 140 (Entrerios). **Tænioptera variegata,** *Burm. P. Z. S.* 1868, p. 635 (Buenos Ayres); *Hudson, P. Z. S.* 1870, pp. 333, 545, et 1871, p. 260 (Buenos Ayres).

Description.—Above and below smoky grey, clearer on the head and breast; belly, crissum, and under wing-coverts bright rufous; wings black, inner secondaries bright chestnut terminated with white, outer secondaries black, tipped with white; wing-coverts grey, margined with white; tail black, outer margins of external pair of rectrices and tips of all whitish; two outer primaries emarginated; bill and feet black: whole length 9·5 inches, wing 6·7, tail 3·8. *Female* similar, but outer primaries not emarginated.

Hab. Patagonia, Argentina, and Paraguay.

There is a striking resemblance to a Thrush in this species, when one sees it running on the ground with its beak somewhat elevated; but when it stands or perches, opening and closing its broad tail with a graceful fan-like motion, the resemblance to the stiff automatic *Turdus* grows less, and when it flies vanishes altogether—its long wings being as sharply pointed as those of the Peregrine Falcon, while its motions in the air have a Gull-like grace and buoyancy.

It is a very pretty bird; the upper plumage is grey tinged with rufous, the throat pure dark grey, breast and belly rufous, wing-coverts light silvery grey, remiges and rectrices dark. Azara classed it under the name of *Pepoaza* (banded-wing) with the *Tæniopteræ*, to which it comes very near in form, flight, language, and habits, though it has longer legs and runs more on the ground. Its summer home is in Southern Patagonia, but its breeding-habits are not known; in winter it migrates north, and in May is found scattered over the pampas, where it is usually called by the country people 'Chorlo,' a name for all Plovers; for while running swiftly about on the ground, often associating with flocks of Plover, it has a certain resemblance to them. From the hue of its plumage it is also called 'El Chocolate,' a name I have thought it best to preserve.

These birds are very sociable, going in small flocks, usually of from half a dozen to twenty individuals; they are restless and active, and quick and graceful in all their movements, and seek their food on the ground, chiefly coleopterous insects, on the great level plains they inhabit. While on the wing they pursue each other playfully in the air, and also attack and chase passing birds of other kinds, apparently in a sportive spirit. Occasionally they perch on a thistle-top or low bush, but never on trees. Their only language is a long, low, plaintive whistle, heard usually on warm still days in winter.

---

112. TÆNIOPTERA NENGETA (Linn.).

(PEPOAZA TYRANT.)

Tænioptera nengeta, *Burm. La-Plata Reise*, ii. p. 459 (Paraná); *Scl. et Salv. Nomencl.* p. 42; *Barrows, Bull. Nutt. Orn. Cl.* viii. p. 137 (Entrerios); *White, P. Z. S.* 1882, p. 603 (Misiones).

Description.—Above cinereous; lores white; wings black, coverts cinereous; a well-marked speculum at the base of the primaries and the edgings of the outer secondaries white; tail black, tipped with whitish cinereous, basal one-third of tail white: below pale cinereous, middle of throat white, with blackish stripe on each side; middle of belly, flanks, crissum, and under tail-coverts white; bill horn-colour; feet black: whole length 9·0 inches, wing 5·0, tail 3·5. *Female* similar.

Hab. S.E. Brazil, Paraguay, Uruguay, Argentine Republic, and Bolivia.

To this species Azara gives the name of *Pepoazá*, the Guarani for Barred-wing; and *Pepoaza* was used by him as a generic name for the small, well-defined group now placed in the genus *Tænioptera*, comprising eight known species. Most of these birds have some conspicuous wing-mark. They inhabit the southern portion of the South-American continent, from South Brazil and Bolivia to the Straits of Magellan, and are most numerous on the open pampas and in Patagonia. In size they do not vary greatly, the largest being about nine inches long, the smallest about seven. In colour they are grey, or, more frequently, white relieved with black or grey, one species (*T. rubetra*) being rufous. Their legs are long, and they run on the ground like *Myiotheretes rufiventris*, feeding, to some extent, in the same manner; but they also occasionally pursue and capture insects on the wing, like the typical Tyrant-birds that seldom or never alight on the ground. They have likewise another and unique preying-habit, intermediate between the Plover-like habits of *Agriornis*, *Myiotheretes*, and *Muscisaxicola*, and the Swallow- or Flycatcher-like habits of the true Tyrants. The bird perches itself on an elevation—the summit of a stalk, or bush, or even of a low tree—to watch like a Flycatcher for its insect prey; only instead of looking about for passing insects, it gazes intently down at the ground, just as a Kingfisher does at the water, and when it spies a beetle or grasshopper, darts down upon it, not, however, to snatch it up with the bill as other Tyrants do, but it first grasps it with its feet, then proceeds to despatch it, swaying about and opening its wings to keep its own balance, just as an Owl is seen to do when it grasps a mouse or other small animal in its claws. After devouring the insect on the spot, it flies back to its perch to resume the watch. They are very restless, active, playful birds, and seldom remain long on one spot, apparently finding it irksome to do so; but I have seen the *T. irupero* occupy the same perch for hours every day while looking out for insects.

As an English generic name for this small interesting group might be useful, I would suggest *Ground-gazers* or *Ground-watchers*, which describes the peculiar preying-habit of these birds.

The Pepoazá is a swift, active, graceful bird, with a strong, straight beak, hooked at the point, and a broad tail four inches long, the total length of the bird being nine inches. The throat and space between the beak and eye are white; all the rest of the body, also the wing- and tail-coverts, light grey; tail and wing-quills black, with a pure white band across the base of the primaries. The tertiaries and rectrices are tipped with pale rufous grey.

It inhabits Brazil south of the equator, Bolivia and Paraguay, also the northern provinces of the Argentine Republic. Mr. Barrows gives the following account of its lively habits in Entrerios:—"They are commonly seen perched on fences or the tops of bushes or trees in open ground, frequently making sallies for winged insects, or dropping to the ground to catch a grasshopper or worm. When shot at while perched and watching you, they almost invariably leave the perch at the flash, pitching forward and downward, and usually evading the shot, even at short range. Several times I have secured them by shooting about a foot below and two feet in front of them as they sat, but they do not always fly in this direction. The rapidity of their flight when frightened, or when quarrelling, is simply astonishing. I have seen one chase another for three or four minutes, doubling, turning, twisting, and shooting, now brushing the grass, and now rising to a height of at least two or three hundred feet, and all the movements so rapid that the eye could scarcely follow them; and at the end of it each would go back to the top of his own chosen weed-stalk, apparently without a feather ruffled."

Azara found this species breeding in a hole in a bank; and Mr. Dalgleish has described a nest, taken from a tree in Uruguay, as a somewhat slight structure, four inches in diameter, formed of sticks and fibres, lined with fine grass and a few feathers. It contained three eggs, pear-shaped, white, with large well-defined spots of reddish brown.

---

113. TÆNIOPTERA CORONATA (Vieill.).

(BLACK-CROWNED TYRANT.)

Tænioptera coronata, *Burm. La-Plata Reise*, ii. p. 459; *Scl. et Salv. Nomencl.* p. 42; *Durnford, Ibis*, 1877, p. 176 (Buenos Ayres); *White, P. Z. S.* 1882,

p. 603 (Buenos Ayres); *Döring, Exp. al Rio Negro, Zool.* p. 42 (R. Colorado); *Barrows, Bull. Nutt. Orn. Cl.* vol. viii. p. 138 (Entrerios). **Xolmis variegata,** *Darw. Zool. Voy. Beagle,* iii. p. 54 (Santa Fé).

Description.—Above cinereous; rounded summit of head black, broad front and band encircling the black of the head white; wings blackish, upper coverts cinereous, edgings of middle and greater coverts and of outer secondaries whitish; tail blackish, margins of outer webs of external tail-feathers white: beneath white; under wing-coverts and a large portion of the inner webs of the remiges, except of the two outer primaries, white; bill and feet black: whole length 7·8 inches, wing 4·6, tail 3·1. *Female* similar.

Hab. Argentine Republic, Uruguay, and Paraguay.

In this species the sexes are alike. The crown is black and composed of loose feathers; the forehead, and a broad line over the eye which extends nearly round the head, also all the under plumage, pure white; neck and back clear grey; quills black.

This Tyrant is a solitary bird, though often many individuals are found within call of each other, and they sometimes even unite in a loose flock. It is found throughout the Argentine country, ranging south to the Rio Negro, in Patagonia, but abounds most on the Buenos-Ayrean pampas, where it performs a partial migration. Most of the *Tænioptera* seek their food by preference on the bare level ground, or where the vegetation is most scanty. This species varies somewhat in habits, and seldom runs on the ground, and chiefly inhabits the desert plains, where the large grasses flourish. On one occasion when I was with an expedition on the pampas for several weeks, every day a number of these birds would gather and follow us; perched here and there on the tall grasses with their bosoms toward us, they often looked at a distance like large white flowers. Old gauchos have told me that fifty years ago they were abundant all over the pampas, but have disappeared wherever the giant grasses have been eaten down, and have given place to a different vegetation.

Their note is a long, low whistle, the usual language of the *Tænioptera*; but in this species it is very like a human whistle, on account of which the bird is named *Boyero* (ox-driver) on the pampas. One severe winter great numbers of them appeared in the neighbourhood of Buenos Ayres, and it was amusing to see the dogs thrown into a great state of excitement by the low whistling notes heard perpetually from all sides. Every few moments they would start up and stare about them to ascertain where the deceptive call came from, and in spite of many disappointments they would occasionally all rush away, loudly barking, into the plantation, convinced that some person there was whistling to call them.

The Black-Crown makes a somewhat shallow nest in a bush or large clump of grass, and lays four white eggs, with large dark red spots, chiefly at the big end.

I cannot refrain from quoting a passage from Mr. Barrows's paper, descriptive of the lively temper and habits of this bird :—

"This species frequently persecutes smaller birds in a way which seems to imply pure love of mischief. One afternoon in July, when the river had fallen some feet after an unusual rise, I was walking along the lines of drift left by the falling water, and watching the different birds which were picking up insects or other food from the wind-rows. A score or two of the little chestnut-backed *Centrites* were running about, and here and there a *Tænioptera* was looking quietly on. Suddenly I heard a chirp of distress, and looking up saw one of these small birds apparently making every effort to escape from a *Tænioptera*, which was following in full chase. The two birds were hardly a length apart and both going at full speed, doubling and dodging in a way that would have done credit to a bat. The chase lasted perhaps for half a minute, when the smaller bird alighted, and at once the other also alighted and began running about unconcernedly and picking up food. But the instant the smaller one made a start his enemy was at his heels (or more properly his tail) again, and he was forced to alight. This was repeated so often that I was on the point of shooting the pursuer, when, without any notice, he flew quietly off, and resumed his usual demeanour. It looked like a case of simple spite, for even if there were twenty other birds about, one seemed to be followed without regard to the rest."

I have often watched *Tænioptera* of different species, also *Myiotheretes rufiventris*, behaving in a similar way, and agree with Mr. Barrows that it is "an *amusement* in which the larger bird indulges simply for the pleasure derived from the exercise of his power."

---

114. TÆNIOPTERA DOMINICANA (Vieill.).
(DOMINICAN TYRANT.)

Tænioptera dominicana, *Burm. La-Plata Reise,* ii. p. 460 (Entrerios); *Scl. et Salv. Nomencl.* p. 42; *Durnford, Ibis,* 1877, p. 176 (Buenos Ayres); *White, P. Z. S.* 1883, p. 38 (Buenos Ayres); *Döring, Exp. al Rio Negro, Zool.* p. 42 (Azul, Currumalan); *Barrows, Bull. Nutt. Orn. Cl.* vol. viii. p. 139 (Entrerios, Pigué, Pampas).

Description.—Above pure white; wings black, with a broad whitish subapical band across the first six primaries, beyond which the tips are blackish; tail black; beneath pure white: whole length 8·0 inches, wing 4·6, tail 4·2, *Female* similar, but head above and back cinereous.

Hab. S.E. Brazil, Paraguay, Uruguay, and Argentine Republic.

This bird ranges from South Brazil and Paraguay to the southernmost pampas of Buenos Ayres. Its total length is eight inches. The wings and tail are black, the former barred with white; all the rest of the plumage in the male is pure white; in the female the upper parts are grey.

It is to some extent migratory, and usually goes in flocks of a dozen or twenty birds, and frequents open situations where there are bushes and trees, also plains covered with giant grasses. They are more social in their habits than *T. coronata*, but in other respects closely resemble it, and are exceedingly active lively birds, and when the flock is on the wing continually pursue each other in a playful manner.

Mr. Barrows observed them in autumn on the Pigué (southern pampas) preparing for their migration. "Late in March," he says, "we found them in large scattered flocks, which collected in one place toward evening, and went through a series of aerial evolutions, accompanied with vocal exercises of a varied and entertaining kind, lasting half an hour or more.

"I presume this was in preparation for their northward (or westward?) migration, as we did not see them again after leaving this spot."

---

115. TÆNIOPTERA IRUPERO (Vieill.).

(WIDOW TYRANT.)

Tænioptera irupero, *Scl. et Salv. P. Z. S.* 1868, p. 141 (Buenos Ayres); *iid. Nomencl.* p. 42; *White, P. Z. S.* 1882, p. 603 (Catamarca, Misiones); *Döring, Exp. al Rio Negro, Zool.* p. 42 (R. Colorado); *Barrows, Bull. Nutt. Orn. Cl.* vol. viii. p. 139 (Entrerios). **Tænioptera mœsta,** *Burm. La-Plata Reise,* ii. p. 460.

Description.—Above and beneath pure white; wings with the primaries black except the innermost, which are white at their bases and tipped with black, and secondaries which have narrow black shafts; broad end of the tail black; bill and feet black; two outer primaries acuminated: whole length 7·0 inches, wing 4·3, tail 3·2. *Female* similar.

Hab. Paraguay, Uruguay, Argentine Republic, and Bolivia.

This pretty species is found throughout the Argentine country, and is well known to the natives, and usually called *Viudita* (Little Widow), on account of its mourning colours. It is also sometimes curiously named *Anjelito de las Animas,* from a superstitious notion due to the intense whiteness of its plumage and to its supposed habit of frequenting graveyards. In both sexes the entire plumage is snowy white, except the primaries and the tip of the tail, which are black. In habits it is more sedentary than other *Tæniopteræ*, and obtains its food chiefly by patiently watching the surface of the ground for its insect prey. Its marvellously white plumage, and the habit of sitting motionless on the summit of a bush or tree, make it a most conspicuous object, so that it is strange to find such a bird existing in districts which abound in raptorial species; for Hawks, I have frequently noticed, will always single out a white or conspicuously coloured bird for pursuit, and though the Little Widow, like the other members of its genus, is swift and strong of wing, the feeble and the young must often fall victims to their shining white plumage.

The Little Widow is a solitary bird, and not nearly so lively and playful in manner as *T. coronata* and *T. dominicana,* its surpassing whiteness being its most interesting feature. Its nesting-habits are unlike those of other *Tæniopteræ,* for it breeds only in holes, usually in the bole or branch of a tree; but sometimes it takes possession of the oven of *Furnarius rufus* to lay in. The nest is composed chiefly of feathers and contains four eggs, creamy white, with a few very minute red spots, irregularly distributed. Mr. Dalgleish says, "Some eggs have only two or three spots, none have more than eight or ten."

Mr. Barrows says:—"The adults have several of the primaries remarkably attenuated. Young birds appear to acquire these attenuate primaries only after a complete moult. But I took one specimen which showed one or more primaries with tips of ordinary shape but with a line apparently *worn* into the vane of the inner web, so as to mark out distinctly the attenuate tip, and it seemed as if a little more wearing would cut out a piece which would leave the primary as in the old bird."

---

116. TÆNIOPTERA MURINA (d'Orb. et Lafr.).

(MOUSE-BROWN TYRANT.)

Tænioptera murina, *Hudson, P. Z. S.* 1872, p. 541 (Rio Negro); *Scl. et Salv. Nomencl.* p. 42; *White, P. Z. S.* 1883, p. 38 (Cordova). **Pyrope murina,** *Cab. Journ. f. Orn.* 1878, p. 196 (Cordova).

Description.—Above sandy cinereous, whitish round the eyes; wings and tail blackish with whitish edgings; below much paler, throat whitish with slight black striations; belly and crissum tinged with ochraceous; under wing-coverts and flanks pale ochraceous; bill horn-colour; feet black; two outer primaries acuminated: whole length 7·0 inches, wing 4·0, tail 2·9. *Female* similar, but outer primaries normal.

Hab. Western Argentina and North Patagonia.

This species inhabits the Mendoza district, and migrates south in spring. I met with it on the Rio Negro, in Patagonia, where it made its appearance in October. The sexes are alike. The entire upper plumage is dull grey with a pale rufous tinge; throat, breast, and belly pale buff tinged with grey. It is a solitary bird, restless in manner, has a swift flight, and sits on a stalk or other slight elevation, from which it darts down to seize any insect it spies on the ground. Its only language is a very low whistling note.

---

117. TÆNIOPTERA RUBETRA (Burm.).

(CHAT-LIKE TYRANT.)

[PLATE VII.]

Tænioptera rubetra, *Burm. La-Plata Reise,* ii. p. 461 (Mendoza); *Hudson, P. Z. S.* 1872, p. 541 (Rio Negro); *Scl. et Salv. Nomencl.* p. 42; *Durnford, Ibis,* 1877, p. 34, et 1878, p. 394 (Patagonia).

Description.—Above sandy brown, lores and superciliaries white; wings black, greater coverts and outer secondaries edged with whitish, lesser coverts like the back; tail black, outer web of the outer tail-feathers and tips of others white; below white, with black striations on the sides of the throat and on the breast; flanks, under wing-coverts, and inner webs of the primaries deep rufous; two outer primaries acuminated: whole length 7·5 inches, wing 4·7, tail 3·2. *Female* rather paler, throat and breast washed with ochraceous, and outer primaries not acuminated.

Hab. Interior of Argentine Republic and Patagonia.

I have met with this bird at all seasons of the year in Patagonia on the Rio Negro, and think it probable that it has no migration. It is seen in flocks of twenty or thirty individuals, and in its lively actions when on the wing, and in its habit of perching on a bush or elevation of some kind, from which it pounces down on an insect seen on the ground, it resembles other *Tæniopteræ*; but it runs about on the ground a great deal, and in this respect is more like a *Myiotheretes* or *Muscisaxicola*. In its colour it also diverges widely from the typical *Tæniopteras* in their black and white Dominican plumage. The whole upper parts are light chestnut, with a white mark on the side of the head; wings and tail dark, tipped with pale rufous; throat, breast, and belly whitish rufous, with dark lines on throat and bosom. The chestnut hue in the female is paler and mixed with grey.

TÆNIOPTERA RUBETRA.

118. OCHTHŒCA LEUCOPHRYS (d'Orb. et Lafr.).

(WHITE-BROWED TYRANT.)

Ochthœca leucophrys, *Scl. et Salv. Nomencl.* p. 42; *White, P. Z. S.* 1882, p. 603 (Catamarca).

Description.—Above dark cinereous, rump rufescent; wings blackish, cross-bands on wing-coverts and edges of outer secondaries rufous; tail blackish, outer web of external rectrix white; beneath pale cinereous; lower belly, crissum, and under wing-coverts white; bill and feet black: whole length 5·2 inches, wing 3·0, tail 2·7.

Hab. Bolivia and Northern Argentina.

A single specimen of this bird was obtained by White at Fuerte de Andalgala, Catamarca, in September 1880, during a snow-storm.

---

119. SAYORNIS CINERACEA (Lafr.).

(ASHY TYRANT.)

Sayornis cineracea, *Scl. et Salv. Nomencl.* p. 43. **Aulanax latirostris,** *Cab. J. f. O.* 1879, p. 335 (Tucuman).

Description.—Above dark cinereous; head sooty black; wings and tail blackish, outer margins of the wing-coverts and secondaries and outer web of the external tail-feather white; below sooty black, middle of belly and under wing-coverts white, flanks and crissum dark cinereous; bill and feet black: whole length 9·5 inches, wing 3·4, tail 3·1. *Female* similar.

Hab. Colombia, Venezuela, Ecuador, Peru, Bolivia, and N. Argentina.

Examples of this wide-ranging species, obtained by Döring in Tucuman, are referred by Dr. Cabanis to his subspecies "*latirostris,*" which seems to us hardly distinct from *S. cineracea.*

---

120. FLUVICOLA ALBIVENTRIS (Spix).

(WHITE-BELLIED TYRANT.)

Fluvicola albiventris, *Scl. et Salv. Nomencl.* p. 43; *Durnford, Ibis,* 1878, p. 59 (Buenos Ayres).

Description.—Above black; front half of head, narrow band across the rump, and slight edgings to wing-coverts and outer secondaries white; below white; bill and feet black: whole length 5·5 inches, wing 2·8, tail 2·2. *Female* similar.

Hab. Amazonia, Bolivia, and Argentine Republic.

The small black-and-white Tyrant is not uncommon in the marshes and on the river-margins in the Plata district, its spring migration extending south to Buenos Ayres. Like the Kingfisher, it haunts the water-side and is found nowhere else. It has a shy, retiring disposition, concealing itself in the close thickets overhanging a stream, so that one does not often see it, notwithstanding its conspicuous white plumage. When disturbed it emits a series of low ticking notes, or darts swiftly out from the thicket, showing itself for a moment over the water before disappearing once more into its hiding-place.

D'Orbigny says it makes a purse-shaped nest, of slender twigs, moss, and feathers neatly interlaced, and lays four white eggs, spotted at the large end with brown.

---

121. ARUNDINICOLA LEUCOCEPHALA (Linn.).

(WHITE-HEADED TYRANT.)

Arundinicola leucocephala, *d'Orb. Voy., Ois.* p. 334 (Corrientes).

Description.—Black; whole head and neck and a patch on the flanks white; bill horn-colour, base of lower mandible white; feet black; whole length 5·0 inches, wing 2·5, tail 1·8. *Female* above cinereous; front and sides of head whitish; tail black; beneath white, flanks and under wing-coverts cinerascent.

Hab. Colombia and southwards to Argentina.

This species, which is of wide distribution, was met with in Corrientes by d'Orbigny.

---

122. ALECTRURUS TRICOLOR, Vieill.

(COCK-TAILED TYRANT.)

Alectorurus tricolor, *Scl. et Salv. Nomencl.* p. 43.

Description.—Above black, rump greyish; sides of the head, scapularies, lesser wing-coverts, and outer margins of secondaries white; tail black, outer rectrix on each side produced, expanded, fan-shaped; below white, patch on each side of the breast (forming an incomplete collar) black; bill horn-colour; feet black: whole length 7·2 inches, wing 2·8; tail, outer rectr. 2·5, middle rectr. 1·5. *Female*: above brown, rump and lesser wing-coverts pale; beneath dirty white, sides of breast brown.

Hab. S. Brazil, Paraguay, Uruguay, and Argentine Republic.

This species generally resembles the one next described, and has, like it, a black, white, and grey plumage. But the tail, although strange, is constructed on a different pattern. The total length of the bird is five and a half inches, the tail being only two and a half. The two outer tail-feathers have remarkably stout shafts, with broad coarse webs, and look like stumps of two large feathers originally intended for a bigger bird, and finally cut off near their base and given to a very small one. In the male these two feathers are carried vertically and at right angles to the plane of the body, giving the bird a resemblance to a diminutive cock; hence the vernacular name 'Gallito,' or Little Cock, by which it is known.

I have not observed this species myself, but Azara has the following paragraph about its habits:—"The male sometimes rises slowly and almost vertically, with tail raised, and rapidly beating its wings, and looking while ascending in this way more like a butterfly than a bird; and when it has reached a height of ten or twelve yards, it drops obliquely to the earth and perches on a stalk." He adds that the males are solitary, but several females are sometimes seen near together, and that the females are greatly in excess of the males.

---

123. ALECTRURUS RISORIUS (Vieill.).

(STRANGE-TAILED TYRANT.)

Alectrurus guira-yetapa, *Scl. et Salv. Nomencl.* p. 43; *Durnford, Ibis,* 1878, p. 60 (Buenos Ayres). **Alectrurus risorius,** *Barrows, Bull. Nutt. Orn. Cl.* viii. p. 140 (Entrerios). **Alectrurus psalurus,** *Burm. La-Plata Reise,* ii. p. 457 (S. Luis).

Description.—Above black, rump grey; front varied with white; wings black, scapularies, outer margins of wing-feathers and coverts white; tail black, two outer rectrices much elongated, denuded at the base, with a broad inner and no outer vane; below white, broad band across the breast black; throat in the breeding-season bare of feathers and of a bright orange; bill yellowish; feet black: whole length 11·0 inches, wing 3·0; tail, outer rectrices 8·0, middle 2·0. *Female*: above brown, wings varied with white; beneath white; breast-band pale brown; tail with the two outer rectrices slightly elongated and denuded, terminated with spatulations on the inner vane.

Hab. S. Brazil, Paraguay, Uruguay, and Argentine Republic.

Azara named this species *Cola estraña* (Strange-tail), but mentions incidentally that its Guarani name is 'guira-yetapá' (Scissor-tail), a term which the Indians apply indiscriminately to several species having the same sort of tail.

The Guira-yetapá is a very curious little bird, with a black, white, and grey plumage and the beak of a true Tyrant; but it differs from all its congeners in having the skin of the chin, throat, and sides of the head bare of feathers, and these parts in the breeding-season are a bright orange colour. It is a feeble flier, its wings being very short, while the two outer tail-feathers are abnormally long and peculiar in form. Mr. Barrows says:—"The remarkable condition of the outer pair of tail-feathers is interesting. In the male these two feathers reach a length of nearly ten inches, the rest of the tail being about three inches in length. The vane on the *inner* side of each is wanting for the first two inches, and then suddenly develops to a width of nearly two inches, which it maintains almost to the tip, when it gradually narrows. The vane on the *outer* side of the shaft is only about one-quarter of an inch wide, and is folded so tightly against the shaft that it is quite inconspicuous. In the only two males of this species which I have seen flying, these long feathers seemed to be carried folded together *beneath* the rest of the tail, and stretching out behind like a rudder or steering-oar, their vanes at right angles to the plane of the rest of the tail."

Mr. Gibson gives a different account, and says the flight is singularly feeble, resembling the fluttering passage of a butterfly through the air, while the tail streams out behind.

It inhabits Paraguay, Uruguay, and the eastern portion of the Argentine Republic, ranging as far south as the pampas in the neighbourhood of Patagonia. It is usually seen singly or in pairs; Azara says he saw a flock of thirty individuals, but as they were all *females*, it may be that in this species, as in *Lichenops perspicillata*, the females are sometimes gregarious, and the males always solitary. It frequents open places, such as the borders of marshes, or plains covered with tall grasses, and perches in a conspicuous place, from which it darts at passing insects like a Flycatcher.

Mr. Gibson found its nest on the ground amongst herbage, and describes it as a neat structure of dried grass, containing three white eggs with a faint cream-coloured tinge.

---

124. CYBERNETES YETAPA (Vieill.).

(YETAPA TYRANT.)

Cybernetes yetapa, *Scl. et Salv. Nomencl.* p. 43; *White, P. Z. S.* 1882, p. 604 (Itapua, Misiones).

Description.—Above hoary grey, with lineiform blackish shaft-stripes; wings black, with a large chestnut-brown patch occupying the outer webs of the inner primaries; tail of twelve much graduated feathers, black; outer web of external rectrices white; beneath same as above, but rather paler; throat and crissum white; patch on each side of the neck, and collar across the neck dark chestnut-brown; under wing-coverts and inner webs of wing-feathers white; bill pale brown; feet black: whole length 16·0 inches, wing 5·0; tail, extreme rectrix 12·5, middle 2·6. *Female* similar, but less bright.

Hab. S.E. Brazil, Paraguay, Bolivia, and Northern Argentina.

Azara found this singular species not uncommon in Paraguay, and since his brief notice of it in the 'Apuntamientos' nothing concerning its habits has been recorded. It has a somewhat laboured flight, he says; lives in pairs or families, frequenting low or marshy grounds, where it perches on the summit of the reeds or bushes, and flies down to the ground to capture insects. It also occasionally dashes after passing insects, taking them on the wing. It has a loud whistle, which it utters frequently without any variation.

---

125. SISOPYGIS ICTEROPHRYS (Vieill.).

(YELLOW-BROWED TYRANT.)

Sisopygis icterophrys, *Hudson, P. Z. S.* 1868, p. 141; *Scl. et Salv. Nomencl.* p. 43; *Durnford, Ibis,* 1877, p. 176 (Buenos Ayres); *Gibson, Ibis,* 1880, p. 29 (Buenos Ayres); *White, P. Z. S.* 1882, p. 604 (Buenos Ayres); *Barrows, Bull. Nutt. Orn. Cl.* viii. p. 141 (Entrerios). **Tæniopterа icterophrys,** *Burm. La-Plata Reise,* ii. p. 460 (Paraná).

Description.—Above bright olive-green, head rather greyish, lores and superciliary stripes yellow; wings blackish, broad ends of coverts and outer edges of secondaries dirty white; tail blackish; beneath bright yellow, sides of breast and flanks olivaceous; under wing-coverts whitish; bill dark horn-colour; feet black: whole length 6·1 inches, wing 3·5, tail 2·7. *Female* similar, but less bright.

Hab. S.E. Brazil, Paraguay, Uruguay, Northern La Plata, and highlands of Bolivia and Peru.

This small and pretty Tyrant-bird is quite common in the woods along the Plata, and is also seen a great deal in orchards and groves in the cultivated districts. In Buenos Ayres it is a summer visitor, appearing there in October, and is a shy, solitary bird, which catches insects on the wing, and rarely visits the ground.

The nest is placed in a tree, ill-concealed, and very shallow: it is built of fine sticks, and lined with fine grass, horse-hair, and feathers. The eggs are four, pointed, pale cream-colour, with large dark red spots, chiefly at the larger end.

The only language of this species is a very low plaintive whistle, uttered as a faint protest when the nest is approached.

The upper plumage is olive-green; the entire under surface and a stripe on the side of the head pure yellow; wing and tail-quills dark. Total length $6\frac{1}{4}$ inches.

---

126. CNIPOLEGUS ANTHRACINUS, Heine.

(ASHY-BLACK TYRANT.)

Cnipolegus anthracinus, *Scl. et Salv. Nomencl.* p. 43; *Salv. Ibis,* 1880, p. 356 (Salta); *Döring, Exp. al Rio Negro, Zool.* p. 24 (Patagonia). **Cnipolegus aterrimus,** *White, P. Z. S.* 1882, p. 604 (Misiones). **Cnipolegus hudsoni,** *White, P. Z. S.* 1883, p. 39 (Cordova)? **Myiarchus fasciatus,** *Leybold, J. f. O.* 1865, p. 402 (Mendoza), ♀.

Description.—Above dull black, a broad bar across the vanes of the inner webs of the wing-feathers white; bill plumbeous; feet black: whole length 6·3 inches, wing 3·5, tail 3·0. *Female* ashy brown: rump, upper tail-coverts, and basal portions of tail bright rufous; wings blackish, with two white transverse stripes; beneath pale fulvous, white on the belly; bill and feet black.

Hab. Northern Argentina.

Unfortunately very little is yet known about the habits of these interesting little Tyrant-birds, for which I should like to suggest the common name "*Spectacular,*" for reasons I shall say more about when I come to describe the *Lichenops perspicillatus,* a species which undoubtedly belongs to this peculiar well-defined group. The plumage of the male is, in most cases, intensely black, and there is a pure white bar on the remiges, hidden when the bird is perched, and when it flies made doubly conspicuous by the peculiar motion of the wings. In all the known species the female has a dull brown plumage, lined or mottled with dusky tints, and with some portion of the wing-quills marked with rufous or chestnut colour.

The Ashy-black Tyrant inhabits the Mendoza district, and is also a summer visitor in Patagonia, where it was obtained by Dr. Döring. Speaking of its habits, he says the male is solitary, perches on the summit of a bush or dry twig, emits at intervals a song or call composed of two syllables, plaintive and flute-like in character, and uttered while the bird rises up a few feet into the air. During this performance the white bands on the wings are displayed conspicuously and a humming sound is produced.

---

127. CNIPOLEGUS HUDSONI, Scl.

(HUDSON'S BLACK TYRANT.)

Cnipolegus hudsoni, *Scl. P. Z. S.* 1872, p. 541, pl. xxxi. (Rio Negro).

Description.—Uniform dull black; a broad bar across the bases of the inner webs of the wing-feathers white; two outer primaries much pointed at their extremities; bill plumbeous, feet black: whole length 6·0 inches, wing 2·8, tail 2·6.

This species is at once distinguishable from the preceding (*C. anthracinus*) by its smaller size and the peculiar narrowed outer remiges.

Hab. Northern Patagonia.

This species is found in the western provinces of the Argentine Republic, and, like *C. anthracinus*, which it closely resembles, is a summer visitor in Patagonia, where it makes its appearance in October. The plumage is intense black, with the inner webs of the remiges at their base white, but the wing-band, which is over an inch in breadth, shows only when the bird flies. There is also a small white spot on the flank, scarcely visible, and excepting for this speck the bird at rest appears entirely black. When it flies the white band appears suddenly, producing a curious effect, for the wings are opened and shut successively and with great rapidity, making the white band appear like a succession of flashes. All the movements of the bird are eccentric to a degree. It selects a dead twig on the summit of a bush, and this perch it occupies during many hours every day. Occasionally it darts after a passing insect, but I believe it feeds principally on the ground, like *Lichenops perspicillatus*. At intervals it quits its perch very suddenly and revolves round it with the rapidity of a moth whirling round the flame of a candle, the wings producing a loud humming sound, and the bird uttering a series of sharp clicking notes. During this performance the white wing-band appears like a pale mist surrounding the bird. This fantastical dance over, it resumes its perch, and, until moved to a second display, sits as motionless as a bird carved out of jet.

---

128. CNIPOLEGUS CYANIROSTRIS (Vieill.).

(BLUE-BILLED TYRANT.)

Cnipolegus cyanirostris, *Burm. La-Plata Reise,* ii. p. 457 (Mendoza); *Scl. et Salv. Nomencl.* p. 43; *Barrows, Bull. Nutt. Orn. Cl.* vol. viii. p. 141 (Entrerios).

Description.—Uniform dull black; inner webs of the wing-feathers margined with dull white; bill plumbeous, feet dark brown: whole length 5·7 inches, wing 3·0, tail 5·6. *Female* dark cinnamomeous brown, brighter on the head and rump; wings blackish, wing-coverts and secondaries edged with pale rufous; tail blackish brown, outer margins of outer rectrices and inner webs of all the rectrices, except the two middle, bright rufous: below pale fulvous, densely striated with blackish; bill and feet brown.

Hab. S. Brazil, Paraguay, Uruguay, and La Plata.

In size this species of Black Tyrant is but slightly smaller than the one described, and the colour is also of an intense black, but the white wing-band is less conspicuous, the borders only of the inner webs of the quills being white. It also differs in having a bright blue beak and red eyes. Azara discovered it in Paraguay, and described the brown-plumaged female as a distinct species. He merely says that it sits on the summit of a bush and darts after passing insects, but it will, no doubt, be found to possess some curious performing habits like the other species of the genus.

Mr. Barrows met with this species on the Lower Uruguay, in the month of September, and Burmeister obtained it near Mendoza.

---

129. CNIPOLEGUS CABANISI, Schulz.

(CABANIS'S TYRANT.)

Cnipolegus cabanisi, *Schulz, Journ. f. Orn.* 1882, p. 462.

Description.—Nearly uniform olivaceous slate-colour; below rather paler; ings and tail black: whole length 6·0 inches, wing 3·4, tail 3·1. *Female* brownish olive, below paler; wings blackish, slightly edged with rufous; tail black, with broad rufous margins to the inner webs, and narrow similar margins to the outer webs; bill and feet brown.

Hab. Tucuman.

This is one of Herr Schulz's discoveries in Tucuman. He met with it, in the month of January, in the mountainous districts of that province.

---

130. CNIPOLEGUS CINEREUS, Scl.

(CINEREOUS TYRANT.)

Cnipolegus cinereus, *Scl. et Salv. Nomencl.* p. 43; *Döring, Journ. f. Orn.* 1878, p. 197 (Cordova); *Salvin, Ibis,* 1880, p. 357, pl. x. (Tucuman, Salta).

Description.—Dark cinereous, paler and more olivaceous on the back and wings; edges of wing-coverts and secondaries rather lighter: below paler, throat and breast blackish, like the head; inner margins of wing-feathers whitish; outer primaries acuminated; bill plumbeous, feet black: whole length 4·6 inches, wing 2·2, tail 2·3. *Female* brownish cinereous; cap and rump rufous; wings blackish, tips of coverts and edges of outer secondaries whitish; tail blackish, inner webs of all the lateral rectrices, except the tips, rufous; beneath whitish.

Hab. Northern Argentina.

---

131. LICHENOPS PERSPICILLATUS (Gm.).

(SILVER-BILL TYRANT.)

Lichenops perspicillatus, *Burm. La-Plata Reise,* ii. p. 457 (La Plata to Mendoza); *Scl. et Salv. P. Z. S.* 1868, p. 141 (Buenos Ayres); *iid. Nomencl.* p. 43; *Hudson, P. Z. S.* 1869, p. 432 (Buenos Ayres); *Gibson, Ibis,* 1880, p. 21 (Buenos Ayres); *White, P. Z. S.* 1882, p. 604 (Buenos Ayres); *Barrows, Bull. Nutt. Orn. Cl.* vol. viii. p. 141 (Entrerios, Azul).

Description.—Uniform black; primaries pure white, with black tips and bases; fleshy ring round the eye and bill yellow, feet black: whole length 5·6 inches, wing 3·5, tail 2·4. *Female* above dark brown, with light brown edgings to the feathers; remiges chestnut, with dark-brown tips; wing-coverts dark brown with fulvous tips; beneath fulvous white, breast with dark striations; under wing-coverts fulvous; bill yellowish, feet black.

Hab. Southern Bolivia, S.E. Brazil, Paraguay, Uruguay, La Plata, Chili, and Patagonia.

Naturalists have said a great deal about the well-known Silver-bill (the most important member of my "*Spectacular*" group), the question as to whether the black and red birds are sexes or two distinct species having long remained unsettled. Azara, writing in the last century, under the heading *Pico de Plata,* rightly described the red bird as the female of the black; but, unfortunately, in another part of his work he described the female again as a different species, naming it *Suiriri chorreado.* Darwin also separated the sexes, and gave the name of *Lichenops erythropterus* to the red-plumaged bird. The following extract gives only a portion of his argument, and is interesting to read, even now, when the question has been finally set at rest, as it shows how much the two birds differ:—"The tail of *L. erythropterus* is squarer and contains only ten feathers instead of twelve: the wing is $\frac{4}{10}$ of an inch shorter; and the secondaries relative to the primaries are also shorter. The red colour on the primaries represents but does not correspond with the white on the black feathers of *L. perspicillatus*; and the secondaries in the two birds are quite differently marked. In *L. erythropterus* the third, fourth, and fifth primaries are the longest, and are equal to each other; the second is only a little shorter than the third. In *L. perspicillatus* the third is rather shorter than the fourth and fifth; and the second is proportionately shorter relative to the third, so that the outer part of the wing in this species is more pointed than in *L. erythropterus.* The outer claw in the latter species is only in a small degree straighter than in the former; and this, considering that *L. perspicillatus* is generally perched, and when on the ground can only hop, and that *L. erythropterus* feeds there entirely, and walks, is very remarkable."

When one considers the habits of the two birds, even where they are most abundant and seen continually, it is indeed difficult to believe that they are of the same species. They are never seen associating together, even in the love-season, and when I have watched a pair actually engaged in constructing their nest, they appeared to keep as far apart as possible. More than that, the male, while unfriendly towards all other species, appears to cherish a special antipathy against the red bird; and when one comes near him never fails to pursue it with the greatest virulence from the neighbourhood. He is also strictly solitary, but the red birds frequently unite in small parties, especially in autumn, when I have often seen as many as a dozen together.

Altogether the Silver-bill has been a puzzle in the past, and it would now appear, from some recent observations made by Mr. Barrows, that we have not yet got to the end of all the curious points in its habits. Without doubt it is migratory. Its range extends from Paraguay to Patagonia, where it is not common. In Paraguay and the hotter parts of the Argentine country it is probably stationary; in Buenos Ayres, where it is most abundant, many individuals remain all the year in sheltered places, and the migration appears to become more definite the further south we get. Mr. Barrows travelled south across the pampas in the autumn, and says:—"The species was met with at all points visited, but south of the Azul not a single male in the black plumage was seen, though the brown birds (presumably females or young) were met with almost every day for nine weeks, and frequently in large numbers. Of course I began to suspect that the males must moult into a brown suit after nesting, as do our Bobolinks and many other birds, but I shot specimens at various times, and all proved to be either females or young males, and as I was confident that at Concepcion black males were to be found through the year, I was at a loss for an explanation, and am so still."

The male Silver-bill is entirely black, there is nothing in nature blacker than its plumage; and, to enhance the effect, the beak is of a very delicate primrose-yellow, which at a little distance appears white, hence the vernacular name. The eye, and broad free skin surrounding it, which is ruffed like an Elizabethan collar, are of the same faint primrose hue. The secondary wing-quills are pure white, but the white is only displayed when the bird flies. The female has the naked skin encircling the eye, but its colour, as also that of the beak, is much darker than in the male. Entire upper plumage dark brown; secondaries chestnut; lower parts fawn-colour, marked with brown. The young males are at first like the females in colour, and do not acquire the black plumage until the end of the summer.

The bird ranges over the whole of the Argentine Republic, and, according to Gay, is also common throughout Chili, where it is known as the *Colegial* (Collegian or learned person), on account of its stiff grave manner, black dress, and spectacled appearance.

The male is a solitary bird, and feeds chiefly on the ground, running rapidly about in open places like a *Muscisaxicola*. It is also frequently seen perched conspicuously on the summit of a tall stalk or bush, and occasionally making a dart into the air after passing insects, showing in this habit his relationship with the Tyrant-birds. But he perches on an elevation less to watch for insects than for the purpose of his curious spectacular performance. This highly eccentric habit is strikingly like that of *Cnipolegus hudsoni*; and I have no doubt that all the *Cnipolegi* possess similar habits. Both birds perch on a conspicuous place, upright, motionless, and looking more like grotesque little automata than living things; they both also leave the perch suddenly, as if shot from it by means of a steel spring. This singularly sudden movement, and the motion of the wings, rapid as in the Humming-bird, or shut and opened alternately and exhibiting the white wing-colour in a series of flashes, seems related to the conspicuous white mark. In both species, also, the wings make a humming sound during flight. The motions of the Silver-bill are, however, in some respects different from those of the *Cnipolegus*. Springing from its perch at intervals, it darts vertically to a height of about fifteen yards, then turns a summersault, uttering at the same moment a shrill-sounding little cry, after which it drops down again and alights on its perch suddenly, as if jerked back to it, and there remains stiff, erect, and motionless as before.

The nest is made of dry grass in a thistle-bush or clump of reeds, and is rather deep and cup-shaped. The eggs are four in number, white, and spotted at the larger end with dark red.

---

132. MACHETORNIS RIXOSA (Vieill.).

(SHORT-WINGED TYRANT.)

Machetornis rixosa, *Burm. La-Plata Reise,* ii. p. 458 (Paraná); *Scl. et Salv. P. Z. S.* 1868, p. 142; *iid. Nomencl.* p. 44; *Hudson, P. Z. S.* 1874, p. 173 (Buenos Ayres); *Durnford, Ibis,* 1877, p. 177 (Buenos Ayres); *Salvin, Ibis,* 1880, p. 357 (Salta); *Barrows, Bull. Nutt. Orn. Cl.* viii. p. 142 (Entrerios).

Description.—Above brownish olive; wings and tail brown, the latter terminated by a yellowish band; middle of cap occupied by a scarlet crest; beneath bright yellow, paler on the throat; bill and feet black: whole length 7·2 inches, wing 3·9, tail 3·5. *Female* similar, but rather duller. Young without the scarlet crest.

Hab. S.E. Brazil, Bolivia, Paraguay, Uruguay, and La Plata.

This species, found in the open districts throughout South America, from Venezuela to Buenos Ayres, where it is quite common, has very interesting habits. It is seven inches and a half long, has a plump body, short wings, and long legs. The upper plumage is light brown, the throat, breast, and belly yellow, and the male has a concealed crest of a bright orange-red colour.

It resembles the true Tyrants in disposition, in its shrill piercing language, and in the habit of perching and breeding in trees. On the other hand, like the long-legged *Myiotheretes*, that lives on the open plains, it feeds exclusively on the ground, over which it runs with a speed possessed by few perching species. The general impression one forms is that in manners and appearance the Short-winged Tyrant is quite unlike any other species, though all its habits are to be found in one or other of the various groups comprising the *Tyrannidæ*.

These birds have no migration, but pair for life, and always remain on the same spot, and will continue to breed in the same hole for many years, even where they are frequently deprived of their eggs. Azara saw them sometimes uniting in small flocks in Paraguay; in Buenos Ayres they are always seen in pairs, or, after the young have left the nest, in families. They prefer to live near a human habitation, where there are trees: even one tree, in which they can breed and find shelter at night, will be sufficient to attach them to a dwelling, so great is their partiality for the clean-trodden ground where they can freely run about and catch insects. They haunt the cattle-pens, and become extremely familiar with the cows, horses, and sheep, following them to the pasture-grounds, where they are often seen perched on the back of a horse or other domestic animal, or stationed close to its nose on the ground, watching for insects. On the bare ground they run about with wonderful swiftness, and are able to overtake and capture flying insects without rising. The male and female invariably hunt together, and at intervals fly to some favourite perch to indulge in a duet composed of loud, rapid, shrill notes, somewhat metallic in sound. Though able to fly swiftly when in pursuit of a passing hawk or other bird, at other times their flight is strangely slow; the round body, short blunt wings and tail giving the bird a somewhat curious appearance as it progresses laboriously through the air. I have frequently seen them make the most unprovoked assaults on birds of an inoffensive kind; possibly they are in these attacks moved by a playful rather than by a vindictive spirit. I once saw one drop like a stone from a height of fifty yards on to a Pigeon perched on a leafless tree. The Pigeon fell as if shot to the earth; the Tyrant-bird then released his hold; the Pigeon rushed away terrified through the trees, while its persecutor rose high up in the air and resumed its journey.

I have elsewhere spoken of the wars waged by this bird against other species, all seeking to gain possession of the large nest of *Anumbius acuticaudatus*. A hole in the trunk of a tree is also a favourite breeding-place. The nest is neatly built of slender twigs and leaves, and lined with horse-hair. The eggs are slightly oval, and densely marked with dark brown spots or stripes on a white or brownish-white ground.

---

133. MUSCISAXICOLA MACLOVIANA (Garn.).

(CHIN-SPOTTED TYRANT.)

Muscisaxicola mentalis, *Hudson, P. Z. S.* 1872, p. 541 (Rio Negro); *Scl. et Salv. Nomencl.* p. 44.

Description.—Above cinereous, lores blackish, cap brown; tail-coverts and tail black, outer margins of outer tail-feather white; below pale cinereous, passing into white on lower belly, crissum, and under wing-coverts; chin-spot brown; bill and feet black: whole length 6·1 inches, wing 4·1, tail 2·6. *Female* similar, but chin-spot not so well marked.

Hab. Peru, Bolivia, Chili, Patagonia, and Falkland Islands.

This South-Patagonian species is one of a Chilian group of Tyrant-birds which resemble in their habits the *Saxicolæ* of Europe. The plumage is generally grey, with more or less rufous colour on the crown; they have long legs, and run swiftly on the ground, frequent open sterile situations, and perch only occasionally on trees.

The present bird is about seven inches long; the upper parts are dull grey, except the crown, which is dark chestnut; under surface light grey, and tail nearly black. In the month of June I met with these birds on the Rio Negro, on their arrival there from the south. They went in flocks of a dozen or twenty birds; they had a swift easy flight, were shy and restless in their manner, and uttered low plaintive whistling notes. When a flock alights on the ground the birds all instantly scatter, running rapidly about in all directions over the bare ground. Occasionally one was seen to perch on some slight elevation, and dart like a Flycatcher after passing insects.

Darwin saw this bird as far north on the Atlantic coast as Bahia Blanca. He also found it at Tierra del Fuego, where it lives entirely on the sea-beaches; and in the sterile upper valleys of the Chilian Andes, at a height of ten thousand feet, where the last traces of vegetation occur and where no other bird lives.

---

134. MUSCISAXICOLA RUFIVERTEX, d'Orb. et Lafr.

(RED-TOPPED TYRANT.)

Ptyonura rufivertex, *Burm. La-Plata Reise,* ii. p. 461 (Mendoza).

Description.—Above clear cinereous; lores and short superciliaries whitish; well-defined nape-patch bright rufous; wings brownish; upper tail-coverts and tail black, outer margin of outer rectrix white; below clear cinereous white, brighter on lower belly, crissum, and under wing-coverts; bill and feet black: whole length 6·5 inches, wing 4·0, tail 2·8. *Female* similar.

Hab. Peru, Bolivia, Chili, and Western Argentina.

---

135. MUSCISAXICOLA MACULIROSTRIS, d'Orb. et Lafr.

(SPOT-BILLED TYRANT.)

Muscisaxicola maculirostris, *Scl. et Salv. Nomencl.* p. 44; *Durnford, Ibis,* 1878, p. 395 (Centr. Patagonia). **Ptyonura maculirostris,** *Burm. La-Plata Reise,* ii. p. 462 (Mendoza).

Description.—Above greyish brown, lores and slight superciliaries whitish; tail-coverts and tail black, the outer web of the outer rectrix white; below pale greyish white, whiter on the belly and crissum; bill black, with the base of the lower mandible yellowish; feet black: whole length 5·5 inches, wing 3·4, tail 2·4. *Female* similar.

Hab. Chili, Patagonia, Bolivia, Peru, and Western Ecuador.

Prof. Burmeister met with this species near Mendoza, in the mouths of the large torrents above the city. Durnford found it near the river Sengalen in Central Patagonia in December 1877.

---

136. CENTRITES NIGER (Bodd.).

(RED-BACKED TYRANT.)

Centrites niger, *Burm. La-Plata Reise,* ii. p. 458; *Scl. et Salv. Nomencl.* p. 44; *Durnford, Ibis,* 1878, p. 395 (Chupat); *White, P. Z. S.* 1882, p. 604 (Buenos Ayres); *Döring, Exp. al Rio Negro, Zool.* p. 42 (R. Colorado); *Barrows, Bull. Nutt. Orn. Cl.* viii. p. 142 (Entrerios, Pampas).

Description.—Above and below deep black: whole of back except the rump and scapularies chestnut; bill and feet black; under wing-coverts and wings below black: whole length 5·0 inches, wing 2·8, tail 1·9. *Female* above brown, back fulvous red; tail black; below ashy brown.

Hab. Patagonia, Chili, and Argentina.

The little Red-backed Tyrant comes nearest to *Muscisaxicola mentalis* in habits, but does not perch on bushes and trees, and is less gregarious than that bird. It is the smallest of all those varied members of the *Tyrannine* family which have abandoned forests and marshes and the pursuit of insects on the wing, to live on the wintry uplands of Patagonia, and on the sterile plains bordering on the Andes.

The male is only five and a quarter inches long. The entire plumage of the male is intensely black, except the back, which is bright chestnut. The inside of the mouth and tongue are vivid orange-yellow. The chestnut colour on the female is pale, the rest of the plumage grey, except the quills, which are dark.

Its summer home is in the southern portion of Patagonia, but its nesting-habits are not known. In March it migrates north, and is very common everywhere on the pampas throughout the winter. They arrive in small parties of three or four, or in little loose flocks of about a dozen individuals, travelling with a swift low flight. Males, females, and young, grey like the last, arrive together; shortly after arriving the young males become mottled with black, and before leaving acquire the adult plumage. They appear to leave in spring all together, but from a note by Durnford it would appear that the males travel in advance of the females. He says:—"Males of this species were common at Chupat throughout September and during the first few days of October. On the 5th of the latter month I observed the first females, which gradually increased in number."

The Little Red-backs inhabit open unsheltered plains, and have so great a predilection for bare ground on which they can run freely about, that on their arrival on the pampas, where the earth is thickly carpeted with grass, they are seen attaching themselves to roads, sheep-pens, borders of streams, vizcacha villages, and similar places. They are exceedingly restless, running swiftly over the ground, occasionally darting into the air in pursuit of small flies, and all the flock so scattered that there will be a dozen yards between every two birds. Mr. Barrows describes their lively habits very well:—"I think this is one of the most restless birds I ever saw. You cannot depend upon him to be in the same place two consecutive half-seconds. He runs like a Sanderling, and whenever he keeps his feet still by accident, his wings are flirted in a way that shows his anxiety to be off. Several are usually found together, and sometimes a loose flock of a hundred or more is seen. They are very strong on the wing, sometimes mounting rapidly for several hundred feet, if suddenly startled, and after a few moments spent in circling like a Snipe, they drop again almost as suddenly as a shot, and as if from the very clouds."

137. PLATYRHYNCHUS MYSTACEUS (Vieill.).

(BROAD-BILLED TYRANT.)

Platyrhynchus mystaceus, *Scl. et Salv. Nomencl.* p. 44; *White, P. Z. S.* 1882, p. 605 (Misiones).

Description.—Above olive, darker on the head; coronal patch bright yellow; lores, eye-region, and ear-coverts pale yellowish; mystacal stripe blackish; wings and tail blackish edged with olive-brown; below clear fulvous, much whiter on the throat; upper mandible blackish, lower whitish; feet pale yellowish: whole length 3·3 inches, wing 2·1, tail 1·1. *Female* similar, but no coronal patch.

Hab. Guiana, Brazil, Paraguay, and Northern La Plata.

A single example of this species was obtained by White in the forest near San Javier, Misiones.

---

138. EUSCARTHMUS MARGARITACEIVENTRIS (d'Orb. et Lafr.).

Todirostrum margaritaceiventer, *d'Orb. Voy., Ois.* p. 316 (Corrientes). **Euscarthmus margaritaceiventris,** *Scl. et Salv. Nomencl.* p. 45; *Salvin, Ibis,* 1880, p. 357 (Tucuman). **Triccus margaritiventris,** *Burm. La-Plata Reise,* ii. p. 456 (Paraná).

Description.—Above olive-green, wings and tail blackish edged with olive-green; whole head above, including sides and back of neck, dark cinereous; beneath pearly white, passing into pale cinereous on the sides; under wing-coverts pale yellowish, flanks tinged with olivaceous; bill hazel; feet red: whole length 4·5 inches, wing 2·1, tail 1·9.

Hab. Paraguay, Northern La Plata, Bolivia, S. Peru, and S.E. Brazil.

This species, discovered by d'Orbigny in Corrientes, was also met with near Paraná by Prof. Burmeister, and by Durnford in Tucuman.

---

139. EUSCARTHMUS GULARIS (Temm.).

(RED-THROATED TYRANT.)

Euscarthmus gularis, *Scl. et Salv. Nomencl.* p. 45; *White, P. Z. S.* 1882, p. 605 (Corrientes and Oran).

Description.—Olive-brown; wings blackish, wing-coverts tipped and outer secondaries edged with yellowish white, forming two distinct bands; remiges and rectrices narrowly margined with olive; head above dark cinereous; lores and sides of the head surrounding the dark ear-coverts rufous; beneath white, sides of the breast greyish; chin and sides of throat rufous, like the lores; flanks and under wing-coverts tinged with yellowish: whole length 3·5 inches, wing 1·8, tail 1·4. *Female* similar.

Hab. S.E. Brazil, Bolivia, and Northern La Plata.

This bird was observed by White near Oran, and also in Misiones.

---

140. PHYLLOSCARTES VENTRALIS (Temm.).

(YELLOW-BELLIED TYRANT.)

Phylloscartes ventralis, *Scl. et Salv. Nomencl.* p. 46; *Barrows, Bull. Nutt. Orn. Cl.* viii. p. 198 (Entrerios).

Description.—Above uniform olive; ill-defined superciliaries whitish; ear-coverts dark; wings and tail blackish with olive margins; well-defined spots on the tips of the two rows of wing-coverts and outer webs of secondaries yellowish; below yellow, rather white on the throat and olivaceous on the sides; under wing-coverts pale yellow; bill and feet blackish: whole length 4·5 inches, wing 2·1, tail 2·3.

Hab. S.E. Brazil and Northern La Plata.

This species is stated to have been met with by Mr. Barrows among the low bushes bordering the streams of Entrerios. Graf v. Berlepsch has recorded its presence in Rio Grande do Sul (Zeitschr. ges. Orn. 1885, p. 131), so that this occurrence is not improbable.

---

141. HAPALOCERCUS FLAVIVENTRIS (d'Orb. et Lafr.).

(REED-TYRANT.)

Arundinicola flaviventris, *d'Orb. Voy. Ois.* p. 335, pl. xxxi. fig. 1. **Hapalocercus flaviventris,** *Burm. La-Plata Reise,* ii. p. 456 (Mendoza); *Scl. et Salv. Nomencl.* p. 46; *Durnford, Ibis,* 1877, p. 177 (Buenos Ayres), et 1878, p. 395 (Centr. Patagonia); *White, P. Z. S.* 1882, p. 605 (Buenos Ayres); *Barrows, Bull. Nutt. Orn. Cl.* viii. p. 199 (Entrerios).

Description.—Above mouse-brown; wings and tail rather darker, with edgings like the back; vertex more or less tinged with rufous; beneath yellow; under wing-coverts pale yellow; bill and feet black: whole length 4·0 inches, wing 2·0, tail 2·0. *Female* similar.

Hab. S. Brazil, Paraguay, Uruguay, La Plata, and Chili.

This little bird is rarely met with in the desert pampas, but throughout the settled portion of the Buenos-Ayrean province it is one of the most common species of the *Tyrannidæ*. It arrives from the north in September, and is very regular in its migrations, although apparently a very feeble flier. It frequents open grounds abounding in thistles, tall weeds, or bushes, and is consequently most abundant about houses. It is extremely active, and occasionally darts after a passing insect and captures it on the wing, especially soft insects, like moths and butterflies, to which it is most partial. It subsists principally, however, on small caterpillars and spiders, for which it searches diligently among the leaves, after the manner of the Wren. Although belonging to the songless division of the Passeres, this small Tyrant-bird possesses a formal song, which the male utters with great frequency, the only other member of the Tyrant-family that I am acquainted with which really sings being the Scarlet Tyrant (*Pyrocephalus rubescens*). The music of the Reed-Tyrant is weak but curious; it is composed of five brief percussive notes, distinctly metallic in sound, which may be imitated by gently and slowly striking *fa la mi sol fa* on the highest keys of the piano. To utter this quaint little song the bird perches itself on the summit of a weed or bush, where it solicits attention with a little chipping prelude, and then jerks its head vigorously with each note, delivering its few drops of sound with all the assurance of a master in the art of melody.

In October it builds a deep elaborate nest of fine dry grass, thistle-down, webs, feathers, and other soft materials, usually in the fork of a weed or thistle three or four feet from the ground. It lays four cream-coloured eggs, the colour deepening to grey at the larger end.

---

142. HABRURA PECTORALIS (Vieill.).

(THIN-TAILED TYRANT.)

Pachyrhamphus minimus, *Gould, Zool. Voy. Beagle*, iii. p. 51, pl. xv. (Monte Video). **Habrura minima,** *Scl. et Salv. Nomencl.* p. 46; *Cab. Journ. f. Orn.* 1878, p. 197 (Cordova); *Scl. P. Z. S.* 1879, p. 460 (Cordova).

Description.—Above sandy brown, with a dark-greyish tinge on the head, which is subcrested and has the vertical feathers white at their bases; lores and eye-region whitish; wings and tail blackish, edged with sandy brown, which forms in some specimens well-marked wing-bands; beneath pale sandy ochraceous, more rufous on the flanks; throat more or less freckled with black; bill and feet dark brown: whole length 4·0 inches, wing 1·9, tail 1·7. *Female* similar, but without the black markings on the throat.

Hab. Northern La Plata, Uruguay, Paraguay, Brazil, and British Guiana.

Examples of this species were obtained by Dr. Döring near Cordova.

---

143. CULICIVORA STENURA (Temm.).

(NARROW-TAILED TYRANT.)

Culicivora stenura, *White, P. Z. S.* 1882, p. 605 (Misiones).

Description.—Above sandy brown, striated with black; head nearly black; lores and superciliaries white; wings and tail blackish with slight brownish edgings; below pale sandy buff, more brownish on the sides of the neck and flanks; bill and feet black: whole length 3·7 inches, wing 1·7; tail, ext. rectr. 1·3, med. 1·9.

Hab. S.E. Brazil.

An example of this species is stated by White to have been obtained by him at Itapua, Misiones, in July 1881, but there was possibly an error in its identification.

---

144. STIGMATURA BUDYTOIDES (d'Orb. et Lafr.).

(WAGTAIL TYRANT.)

Stigmatura budytoides, *Scl. et Salv. Nomencl.* p. 46; *Salvin, Ibis,* 1880, p. 357 (Tucuman, Salta).

Description.—Above greyish olive; lores and short superciliary streak yellowish; wings blackish brown, tips of wing-coverts and outer margins of secondaries white; tail blackish, all the lateral rectrices crossed by a broad, white, median and second terminal band; under surface pale yellow; bill and feet black: whole length 5·0 inches, wing 2·1, tail 2·5.

Hab. Bolivia, Peru, Interior of Brazil, Northern La Plata.

Examples of this species were obtained by Durnford in Tucuman and Salta.

---

145. STIGMATURA FLAVO-CINEREA (Burm.).

(LONG-TAILED TYRANT.)

Phyllcscartes flavo-cinereus, *Burm. La-Plata Reise,* ii. 455 (Mendoza); *Döring, Exp. al Rio Negro, Zool.* p. 42 (R. Negro, R. Colorado). **Stigmatura flavo-cinerea,** *Hudson, P. Z. S.* 1872, p. 542 (Rio Negro); *Scl. et Salv. Nomencl.* p. 46; *White, P. Z. S.* 1882, p. 606 (Catamarca).

Description.—Above greyish olive, lores and superciliary stripe whitish; wings blackish, with whitish edgings to the coverts and outer secondaries; tail blackish; outer web of the external rectrix and broad tips of the four external pairs white; beneath pale yellow; bill and feet black: whole length 5·8 inches, wing 2·3, tail 3·0. *Female* similar.

Hab. Argentine Republic, including N. Patagonia.

This little bird inhabits the Mendoza and Patagonian districts, and does not appear to be migratory, for on the Rio Negro I found it at all seasons. It is slender in form, with a long tail, its total length being six inches. The sexes are alike in colour; the upper parts are yellowish grey, breast and belly light yellow. They are found living in pairs, all the year round, in thorn bushes, and are scarcely ever seen to rest, but hop incessantly from twig to twig, in a delicate, leisurely manner, seeking on the leaves for the minute caterpillars and other insects on which they live. While thus engaged they utter a variety of little chirping and twittering notes, as if conversing together, and occasionally the two birds unite their voices in a shrill, impetuous song.

---

146. SERPOPHAGA SUBCRISTATA (Vieill.).

(SMALL-CRESTED TYRANT.)

Serpophaga subcristata, *Burm. La-Plata Reise,* ii. p. 454 (Entrerios); *Scl. et Salv. Nomencl.* p. 47; *Durnford, Ibis,* 1877, p. 177 (Buenos Ayres); *White, P. Z. S.* 1882, p. 606 (Misiones); *Barrows, Bull. Nutt. Orn. Cl.* vol. viii. p. 199 (Entrerios).

Description.—Above cinereous, usually with a slight olivaceous tinge on the rump; crest-feathers white at their bases, tipped with cinereous, and slightly varied with black; wings blackish, wing-coverts tipped with whitish, forming two handsome bands; outer secondaries externally margined with the same colour; tail dark ashy; beneath ashy white, with more or less yellowish tinge on the belly and under wing-coverts; bill horn-colour; feet black: whole length 4·5 inches, wing 2·0, tail 2·1. *Female* similar.

Hab. S.E. Brazil, Paraguay, Northern La Plata, and Bolivia.

This species is one of the smallest members of our *Tyrannidæ,* its total length being only four and a half inches. The sexes are alike; the upper plumage is grey, with a greenish tinge on the back; the breast paler grey, becoming pale yellow on the belly. There is a white concealed spot under the loose feathers of the crown.

It is quite common in Buenos Ayres, and, probably, has a partial migration, as it is most abundant in summer. In its habits it closely resembles the species last described, being always found in pairs, living in thickets, where they hop incessantly about, exploring the leaves for small caterpillars, and always conversing in low, chirping, and twittering notes. They also sing together a little confused song. The nest is fastened to the slender twigs of a low bush, and is a deep, cup-shaped and beautiful structure, composed of a great variety of soft materials bound together with spiders' webs, the interior lined with feathers or vegetable down, and the outside with lichen. The eggs are two, bluntly pointed, and of a cream-colour.

147. SERPOPHAGA NIGRICANS (Vieill.).

(BLACKISH TYRANT.)

Serpophaga nigricans, *Burm. La-Plata Reise,* ii. 454 (Paraná); *Scl. et Salv. Nomencl.* p. 47; *Durnford, Ibis,* 1877, p. 177 (Buenos Ayres); *White, P. Z. S.* 1882, p. 606 (Misiones); *Barrows, Bull. Nutt. Orn. Cl.* vol. viii. p. 199 (Entrerios).

Description.—Above dull brownish cinereous; wings and tail blackish, the coverts and outer secondaries with slight edgings like the back; crest slight, with a well-marked white basal spot; beneath paler and rather purer cinereous; under wing-coverts pale cinereous; bill and feet dark horn-colour: whole length 4·7 inches, wing 2·3, tail 1·1. *Female* similar, but vertical spot not so well marked.

Hab. S.E. Brazil, Paraguay, La Plata, and N. Patagonia.

This species differs markedly in habits, language, and appearance from the last. In both sexes the colour is a uniform slaty grey; the tail, which the bird incessantly opens and flirts like a fan, is black; as in *S. subcristata* there is a hidden spot of white under the loose feathers forming the crest.

It frequents the borders of running streams, seldom being found far from a water-course; and it alights as often on stones or on the bare ground as on trees. Male and female are always seen together, for it pairs for life, and the migration, if it has any, is only partial. It flits restlessly along the borders of the stream it frequents, making repeated excursions after small winged insects, taking them in the air, or snatching them up from the surface of the water, and frequently returning to the same stand. While thus employed it perpetually utters a loud, complaining *chuck,* and at intervals the two birds meet, and, with crests erect and flirting their wings and tails, utter a series of trills and hurried sharp notes in concert.

The nest is generally placed beneath an overhanging bank, attached to hanging roots or grass, a few inches above the water; but it is sometimes placed in a bush growing on the borders of a stream. It is a neat, cup-shaped, but rather shallow structure, thickly lined inside with feathers. The eggs are four, pointed, white or pale cream-colour, with black and grey spots at the large end.

---

148. ANÆRETES PARULUS (Kittl.).

(TIT-LIKE TYRANT.)

Anæretes parulus, *Burm. La-Plata Reise,* ii. p. 455 (Mendoza); *Hudson, P. Z. S.* 1872, p. 543 (Rio Negro); *Scl. et Salv. Nomencl.* p. 47; *White, P. Z. S.* 1883, p. 39 (Cordova); *Durnford, Ibis,* 1878, p. 395 (Centr. Patagonia); *Döring, Exp. al Rio Negro, Zool.* p. 43 (R. Negro, R. Colorado).

Description.—Above cinereous, with an olivaceous tint on the lower back; head black, front varied with white, elongated vertical crest black, sometimes varied with white; wings blackish, with slight whitish tips to the coverts and whitish margins to the outer secondaries; tail blackish, outer webs of external rectrices whitish; below pale straw-colour, white on the throat; throat and breast with numerous and well-marked black striations; bill and feet black: whole length 4·0 inches, wing 1·9, tail 1·8. *Female* similar, but crest shorter.

Hab. Argentine Republic, including Patagonia, Chili, Bolivia, Peru, and Andes of Ecuador.

This small bird is only four and a half inches long; in both sexes the colour on the upper parts is dull grey, on the throat and breast ash-coloured; the belly pale yellow. It has the distinction of a slender curling Lapwing-like crest, composed of a few narrow, long, black feathers. The eye is white. It is found in the thorny thickets on the dry plains of Mendoza, and is also common in Patagonia. In its habits it closely resembles *Serpophaga subscristata*; lives always in pairs, perpetually moves about in a singularly deliberate manner while searching through the bush for small insects, the two birds always talking together in little chirping notes, and occasionally bursting out into a little shrill duet. It builds a deep neat nest of fine dry grass and lined with feathers, in a low thorn, and lays two white eggs.

---

149. ANÆRETES FLAVIROSTRIS, Scl. et Salv.

(YELLOW-BILLED TIT-TYRANT.)

Anæretes flavirostris, *Scl. et Salv. P. Z. S.* 1876, p. 355; 1879, p. 613.

Description.—Above brownish olive; head black, mixed with white and surmounted by narrow elongated black crest-feathers; wings blackish, outer web of external rectrix whitish; beneath pale stramineous, throat and breast white, densely striated with black; under wing-coverts white; bill yellowish, with a black tip; feet black: whole length 4·2 inches, wing 2·1, tail 2·0.

Hab. Bolivia and N. Argentina.

A specimen of this species was procured at Cosquin, near Cordova, by White.

---

150. CYANOTIS AZARÆ, Naum.

(MANY-COLOURED TYRANT.)

Cyanotis azaræ, *Scl. et Salv. Nomencl.* p. 47; *Durnford, Ibis,* 1877, p. 177 (Buenos Ayres, Centr. Patagonia); *Gibson, Ibis*, 1880, p. 32 (Buenos Ayres); *Döring, Exp. al Rio Negro, Zool.* p. 43 (R. Colorado, R. Negro); *Barrows, Nutt. Bull. Orn. Cl.* viii. p. 200 (Entrerios, Carhué, Pampas).

Description.—Above dark bronzy green; head black; superciliaries yellow; vertical spot crimson; wings black; broad tips of the lesser wing-coverts and broad edgings of some of the secondaries white, forming a large white bar on the wing; tail black, greater part of outer pair of rectrices and outer web and broad tip of next pair and narrow tips of third pair white; beneath bright ochreous yellow; chin whitish; crissum crimson; incomplete band across the lower breast black; under wing-coverts white; bill black; feet dark flesh-colour: whole length 4·2 inches, wing 2·1, tail 1·7. *Female* similar.

Hab. La Plata, Chili, and Western Peru.

This charming little bird is variously called by the country people All-coloured or Seven-coloured. Azara called it "The King"—a name which this species deserves, he says, not only on account of the crown of loose feathers on its head, but because it is exceeded by few birds in beauty. It is the most beautiful bird found in Chili, says Gay; and Darwin, who is seldom moved to express admiration, calls it "an exquisitely beautiful little bird." There are many species possessing a more brilliant plumage, none with so great a variety of distinct colours; for on its minute body, which is less than that of the House-Wren, are seen black, white, green, blue, orange, yellow, and scarlet; and all these hues are disposed and contrasted in such a manner as to produce a very pleasing effect—the olive-green and delicate yellow predominating, while the vivid scarlet is a mere spot, like the bright gem or ornament which serves to set off and enhance the beauty of the dress. The whole under plumage is pure lovely yellow, while a broad mark of velvet-black extends belt-wise from the bend of each wing, but without meeting in the centre of the bosom. The sides of the head are deep blue; over the blue runs a bright yellow stripe, surmounted with the loose, slender, almost hair-like feathers of the crown, which stand partially erect, and are blue mixed with black, with vivid scarlet in the centre. Above, from the back of the head to the tail, the colour is deep green. The wings are black, crossed with a white band; tail also black, the two outer quills pure white, and the succeeding two partially white, the white colour appearing only when the bird flies. Moreover, as though this diversity of colour were not enough, the soles of the black feet are bright orange, the eye of the male delicate sky-blue, while the female has white eyes.

While on the subject of the colouring of this species, I will mention a curious phenomenon which I have observed many times. When the bird is flying away from the spectator in a strong sunlight, and is at a distance of from twenty to thirty yards from him, the upper plumage, which is dark green, sometimes appears bright blue. At first I thought that a distinct species of *Cyanotis*, cerulean blue in colour, existed, but finally became convinced that the green feathers of the *C. azaræ* appear blue in certain lights. This seems strange, as the feathers of the back are not glossed.

The Many-coloured Tyrant is, apparently, a very feeble flier, rising reluctantly when frightened from the rushes, and fluttering away to a distance of a few yards, when it again drops down. Yet it is strictly migratory. Darwin met with it at Maldonado in the month of June, and therefore concluded that it does not migrate; but he mentions that it was very rare. I have also occasionally seen one in winter on the pampas, but many migratory species leave a few stragglers behind in the same way. At the end of September they suddenly appear all over the pampas, in every swamp and stream where there are beds of rushes; for in such situations only is the bird found: and this migration extends far into Patagonia. They are always seen in pairs amongst the dense rushes, where they perch on the smooth stems, not near the summit, but close down to the surface of the water, and perpetually hop from stem to stem, deftly picking up small insects from the surface of the water. They also occasionally leave the rushes and search for insects in the grass and herbage along the border. They are very inquisitive, and if a person approach the rush-bed, they immediately come out of their concealment, both birds uttering their singular notes—a silvery, modulated sound, not meant for a song apparently, and yet I do not know any sweeter, purer sound in nature than this. All through the close-growing dark rushes the pretty little melodists may be heard calling to each other in their delicate, gurgling notes.

The nest is a marvel of skill and beauty. As a rule it is attached to a single polished rush, two or three feet above the water and about the middle of the stem. It is cup-shaped inside, and about four inches long, circular at the top, but compressed at the lower extremity, and ending in a sharp point. It is composed entirely of soft bits of dry yellow rush, cemented together with gum so smoothly that it looks as if made in a mould. The eggs are two in number, oval, and dull creamy white, sometimes with a ring of colour at the large end.

---

151. LEPTOPOGON TRISTIS, Scl. et Salv.

(SORRY TYRANT.)

Leptopogon tristis, *Scl. et Salv. P. Z. S.* 1876, p. 254; *White, P. Z. S.* 1882, p. 606 (Misiones).

Description.—Above olive; cap rather darker; wings and tail blackish, margined with olive, the two rows of wing-coverts and three or four outer secondaries with clear yellowish-white apical spots; beneath pale sulphur-yellow; under wing-coverts yellow; inner margins of wing-feathers whitish; bill horn-colour; feet brown: whole length 4·4 inches, wing 2·1, tail 2·1.

Hab. Northern La Plata and Bolivia.

This bird was obtained by White near San Javier, Misiones, "amongst the lofty trees on the river-banks." White believed that he also met with it in the Sierra de Totoral.

---

152. ELAINEA ALBICEPS (d'Orb. et Lafr.).

(WHITE-CRESTED TYRANT.)

Elainea modesta, *Burm. La-Plata Reise,* ii p. 454. **Elainea albiceps,** *Scl. et Salv. Nomencl.* p. 48; *Durnford, Ibis,* 1878, p. 60 (Buenos Ayres); *Barrows, Bull. Nutt. Orn. Cl.* vol. viii. p. 200 (Entrerios); *White, P. Z. S.* 1882, p. 606 (Buenos Ayres).

Description.—Above dark ashy brown, with a slight olivaceous tinge on the rump; head slightly darker, with a more or less concealed white vertical spot; wings and tail blackish brown; tips of wing-coverts, forming two transverse bands, and outer margins of exterior secondaries dirty white; rest of wing-feathers and tail-feathers slightly margined with lighter colour; below nearly uniform pale cinereous, whiter on the throat and middle of the belly; under wing-coverts slightly tinged with yellowish; bill brownish; feet black: whole length 6·0 inches, wing 3·2, tail 3·0. *Female* similar.

Hab. Peru, Ecuador, Bolivia, Guiana, Brazil, Argentina, Chili, Patagonia.

A widely spread species, very common near Buenos Ayres according to White, where it is found in the clumps of trees.

---

153. ELAINEA STREPERA, Cab.

(NOISY TYRANT.)

Elaïnea strepera, *Cab. J. f. O.* 1883, p. 215.

Description.—Above dark greyish olive; head slightly crested, with a white basal spot; eye-ring white; wings and tail blackish, tips of wing-coverts rufous, slight margins of wing and tail-feathers olivaceous; beneath cinereous; middle of belly white; flanks olivaceous; under wing-coverts pale cinereous; bill dark brown, pale at the base; feet blackish: whole length 5·6 inches, wing 2·9, tail 2·7.

Hab. Tucuman.

Dr. Cabanis established this species, which is unknown to us, on specimens obtained by Herr Schulz in the woods of Tucuman. It is said to have a loud voice, and to feed on berries.

154. ELAINEA VIRIDICATA (Vieill.).

(GREENISH TYRANT.)

Muscicapara viridicata, *d'Orb. Voy., Ois.* p. 325. **Elaïnea grata,** *Cab. J. f. O.* 1883, p. 216 (Tucuman).

Description.—Above dark olive-green; head dark cinereous, slightly crested, with a large basal spot of bright yellow; lores and eye-region mixed with whitish; wings and tail ashy black, with slight margins of the same colour as the back; below pale cinereous; belly, crissum, and under wing-coverts sulphur-yellow; bill blackish; feet dark brown: whole length 5·0 inches, wing 2·5, tail 2·5. *Female* similar.

Hab. South America.

The *Elainea grata,* based by Dr. Cabanis upon specimens obtained by Herr Schulz in Tucuman, must, I think, be identical with Azara's *Contramaestre pardo verdoso, corona amarilla,* upon which Vieillot established his *Sylvia viridicata.* It is certainly, in my opinion, the *Muscicapara viridicata* of d'Orbigny.

Herr Schulz met with this species in the province of Tucuman, in the month of December.

155. EMPIDAGRA SUIRIRI (Vieill.).

(SUIRIRI TYRANT.)

Tænioptera suiriri, *Burm. La-Plata Reise,* ii. p. 460 (Tucuman). **Empidagra suiriri,** *Scl. et Salv. Nomencl.* p. 49; *iid. P. Z. S.* 1869, p. 633 (Buenos Ayres); *Cab. J. f. O.* 1878, p. 197. **Pachyrhamphus albescens,** *Gould, Zool. Beagle,* iii. p. 50, t. xiv. (Buenos Ayres).

Description.—Above cinereous; wings and tail blackish, all the wing-coverts and outer secondaries broadly margined externally with white; outer web of outer tail-feathers white; outer edges of primaries and narrow ends of tail-feathers cinereous; below white, under wing-coverts pale yellowish white; bill and feet black: whole length 5·5 inches, wing 2·9, tail 2·5.

Hab. Argentine Republic, Uruguay, Paraguay, and Bolivia.

This species is stated by Prof. Burmeister to be found in Tucuman and Northern Argentina. It also occurs near Buenos Ayres, where Hudson obtained specimens for the Smithsonian Institution.

156. SUBLEGATUS GRISEOCULARIS, Sclater.

(GREY-EYED TYRANT.)

Sublegatus griseocularis, *Scl. et Salv. P. Z. S.* 1870, p. 17 (Mendoza).

Description.—Above cinereous; wings and tail dark ashy brown, margins of wing-coverts and outer secondaries whitish; below, throat and breast pale cinereous white; belly and under wing-coverts pale lemon-yellow; bill horn-colour; feet dark brown: whole length 5·0 inches, wing 2·8, tail 2·5. *Female* similar.

Hab. Southern Peru and Western Argentina.

Specimens of this species were obtained by Weisshaupt near Mendoza.

---

157. RHYNCHOCYCLUS SULPHURESCENS (Spix).

(SULPHURY TYRANT.)

Rhynchocyclus sulphurescens, *Scl. et Salv. Nomencl.* p. 49; *White, P. Z. S.* 1882, p. 607 (Misiones).

Description.—Above olive-green, head washed with cinereous; lores and line round the eye whitish; wings blackish brown, margined with yellowish olive; tail dark brown, slightly edged with olive-green; below sulphur-yellow tinged with olivaceous, throat more greyish; middle of the belly rather brighter; under wing-coverts pale sulphur-yellow; bill horn-colour; feet brown: whole length 5·2 inches, wing 2·6, tail 2·5. *Female* similar.

Hab. S. America from Northern La Plata to Veragua.

Of this very widely diffused species specimens were obtained by White, in June 1881, in the dense forests of Misiones.

---

158. PITANGUS BOLIVIANUS (Lafr.).

(BIENTEVEO TYRANT.)

Pitangus bellicosus, *Scl. et Salv. Nomencl.* p. 50; *Durnford, Ibis,* 1877, p. 178 (Buenos Ayres); *Gibson, Ibis,* 1880, p. 24 (Buenos Ayres); *White, P. Z. S.* 1882, p. 607 (Catamarca); *Barrows, Bull. Nutt. Orn. Cl.* vol. viii. p. 200 (Entrerios). **Saurophagus sulphuratus,** *Burm. La-Plata Reise,* ii. p. 452.

Description.—Above brown; head black; front, superciliaries, and line round the nape white; large vertical crest yellow, tipped with black; wings and tail brown with rufous margins; beneath sulphur-yellow, inner margins of wing- and tail-feathers pale rufous; bill and feet black: whole length 8·0 inches, wing 4·2, tail 3·3. *Female* similar, but yellow crest not so much developed.

Hab. Bolivia, S. Brazil, Uruguay, and Argentina.

The Bienteveo is in its habits the most interesting member of the Tyrannine family. It would be difficult to find two species more dissimilar in disposition than are the Silver-bill, already described, and the Bienteveo; the former being like an automaton, having only a few set motions, gestures, and instincts, while the other is versatile in an extraordinary degree, and seems to have studied to advantage the various habits of the Kestrel, Flycatcher, Kingfisher, Vulture, and fruit-eating Thrush; and when its weapons prove weak it supplements them with its cunning. How strange it is that these two species, mentally as widely separated as the Humming-bird and Crow, should be members of the same family!

The Bienteveo has a wide range in South America, and inhabits the whole of the Argentine country down to Buenos Ayres, where it is very common. It is resident and lives in pairs, the sexes being always faithful. The body is stout, somewhat large for a Tyrant-bird; the length being nine and a half inches, including the beak, which is a little over an inch in length. The wings are blunt and comparatively short, measuring, when spread, fourteen inches. The head is large, and a broad, black band extends from the beak its entire length, and above this is a pure white stripe; the crown is black, concealing in its loose, abundant feathers a brilliant yellow crest, which shows only when the bird is excited. The upper plumage, including wings and tail, is pale brown; the entire under surface sulphur-yellow. In both sexes the plumage is alike.

In Buenos Ayres the Bienteveo is found in every orchard and plantation; it is familiar with man and invariably greets his approach with loud notes—especially with a powerful three-syllabled cry, in which people fancy there is a resemblance to the words Bien-te-veo ("I see you well"); while its big head and beak, and strongly contrasted colours, especially the black and white head-stripes, seem to give it a wonderfully knowing look, as it turns its head from side to side to examine the intruder. It is a loud-voiced garrulous bird, and has a great range of sounds, from grating screams to long, clear, almost mellow call-notes. It has one pretty habit, which brings out strongly the pleasant feature in its character. Though the male and female are greatly attached, they do not go afield to hunt in company, like the Short-winged Tyrant, but separate to meet again at intervals during the day. One of a couple (say the female) returns to the trees where they are accustomed to meet, and after a time, becoming impatient or anxious at the delay of her consort, utters a very long, clear call-note. He is perhaps a quarter of a mile away, watching for a frog beside a pool, or beating, harrier-like, over a thistle-bed, but he hears the note and presently responds with one of equal power. Then, perhaps, for half an hour, at intervals of half a minute, the birds answer each other, though the powerful call of the one must interfere with his hunting. At length he returns; then the two birds, perched close together, with their yellow bosoms almost touching, crests elevated, and beating the branch with their wings, scream their loudest notes in concert—a confused jubilant noise that rings through the whole plantation. Their joy at meeting is patent, and their action corresponds to the warm embrace of a loving human couple.

I have frequently stood for the space of half an hour concealed amongst the trees where a Bienteveo was calling to her mate, cheered at intervals by the far-off faint response, for the pleasure of witnessing in the end the joyful reunion of the two birds.

Except when breeding the Bienteveo is a peaceful bird, never going out of its way to make gratuitous attacks on individuals of its own or of other species; but in the pursuit of its prey it is cunning, bold, and fierce. Like the true Tyrant-birds it preys a great deal on large insects when they are abundant in the warm season, and is frequently seen catching its prey in the air. A large beetle or grasshopper it invariably beats against a branch before devouring it. But even in summer, when insect prey is most abundant, it prefers a more substantial diet whenever such is to be had. It frequently carries off the fledglings of the smaller birds from their nests, in the face of the brave defence often made by the parents. It is also fond of fishing, and may be seen perched by the hour on a bank or overhanging branch beside a stream, watching the water like a Kingfisher, and at intervals dashing down to capture the small fry. In shallow pools, where there are tadpoles and other prey, the Bienteveo does not mind getting a little wet, but alights in the water and stands belly-deep watching for its prey. I have seen a Bienteveo standing in the water in the midst of a flock of Glossy Ibises. They are often seen, as Darwin remarks, hovering like a Kestrel over the grass, and then dashing down to seize their prey. Small snakes, frogs, mice, and lizards all minister to its appetite, and with a capture of this kind it invariably flies to the nearest stone or branch, against which it beats the life of its victim out before devouring it. I once saw one fly out of some weeds carrying a little wriggling snake of a very brittle species and about eight inches long in its beak. Alighting on a gate it proceeded to kill its capture, and at the first blow on the wood the snake flew into two pieces. A mouse gives it a great deal of trouble, for after it has been killed it cannot be devoured until reduced by repeated blows to a soft pulp, after which it is with great labour pulled to pieces and eaten. Snails and *Ampullariæ* are also pounded until the shell breaks. In spring they sometimes join the train of hooded gulls, guira cuckoos, cow-birds, and various other kinds which follow the plough to pick up worms and larvæ; but on the ground the Bienteveo is awkward in its motions, for it cannot run like the Tyrant-birds with terrestrial habits, but only hops. At estancia houses, when a cow is slaughtered, it comes in with the fowls, milvago hawks, and dogs, for small pickings, being very fond of fresh meat. It is a common thing to see a Bienteveo following a rural butcher's cart, and waiting for an opportunity to dash in and carry off any small piece of meat or fat it is able to detach. In the autumn they feed very much on ripe fruit, preferring grapes, which they can swallow whole, and figs, which are soft and easily eaten.

In its nidification the Bienteveo also departs widely from the, so to speak, traditional habits of its congeners; for whereas most Tyrants make small shallow nests, this species makes a very big elaborate domed structure, and sometimes takes five or six weeks to complete it. It is placed in a tree, without any attempt at concealment, and is about a foot deep and eight or nine inches broad, and composed of a variety of soft materials, chiefly wool. The entrance is placed near the top. Outside, the nest has a very disorderly appearance, as there are always long straws and sometimes rags hanging down; the cavity is deeply lined with feathers, and is the *hottest* nest I know. The eggs are five, very long, pointed, cream-coloured, and spotted, chiefly at the large end, with chocolate and purple.

---

159. MYIODYNASTES SOLITARIUS (Vieill.).

(SOLITARY TYRANT.)

Myiodynastes solitarius, *Scl. et Salv. Nomencl.* p. 50; *Durnford, Ibis,* 1877, p. 178, et 1878, p. 60 (Buenos Ayres); *White, P. Z. S.* 1882, p. 607 (Salta); *Barrows, Bull. Nutt. Orn. Cl.* vol. viii. p. 201 (Entrerios).

Description.—Above grey, thickly covered with numerous black shaft-spots; concealed vertical crest yellow; lores and sides of the head black, superciliaries white; wings blackish, primaries slightly edged with rufous, coverts and secondaries margined externally with greyish white; tail blackish, with narrow external margins of ferruginous; below greyish white, with numerous black striations; belly and under wing-coverts tinged with yellowish and but slightly striated; bill brown, feet blackish: whole length 7·5 inches, wing 4·4, tail 3·4.

Hab. Brazil, Paraguay, Buenos Ayres.

Durnford tells us that this species is a summer visitor to the neighbourhood of Buenos Ayres, but not very common. White met with it in the forests of Campo Santo in November; and Barrows obtained specimens among the Paradise-trees in the Plaza at Concepcion, in the same month of the year.

---

160. HIRUNDINEA BELLICOSA (Vieill.).

(WARLIKE TYRANT.)

Hirundinea bellicosa, *Scl. et Salv. Nomencl.* p. 51; *White, P. Z. S.* 1882, p. 607 (Catamarca). **Hirundinea rupestris,** *Scl. Ibis,* 1869, p. 198, pl. v. fig. 8.

Description.—Above sooty brown; wings blackish, with a large ferruginous red blotch occupying the greater portion of the inner primaries and secondaries; rump and greater part of the tail-feathers ferruginous red; apical portion of tail-feathers blackish: beneath ferruginous red, throat greyish; under wing-coverts and inner webs of wing-feathers, except the tips, similar to the belly, but brighter; bill and feet black: whole length 7·0 inches, wing 4·3, tail 2·2. *Female* similar.

Hab. S.E. Brazil, Paraguay, and Northern Argentina.

This Flycatcher is by no means common in Catamarca. It is seen about the houses in pairs, but not more than one pair at a time at one dwelling, perched upon some projection, whence it darts off into the air at passing insects.

"The snap of its beak, as it dashes at the flies, can be heard a long way off. Its cry is peculiar and piteous. There is no perceptible difference in plumage between the male and female."—*White, l. s. c.*

---

161. MYIOBIUS NÆVIUS (Bodd.).

(LITTLE BROWN TYRANT.)

Myiobius nævius, *Scl. et Salv. Nomencl.* p. 51; *White, P. Z. S.* 1882, p. 607 (Buenos Ayres); *Barrows, Bull. Nutt. Orn. Cl.* vol. viii. p. 201 (Entrerios).

Description.—Above brown; lores whitish; concealed vertical crest red or yellow; wings blackish, two bands across the coverts and outer margins of external secondaries pale rufous or fulvous, sometimes whitish; tail dark brown: beneath fulvous white; sides of throat, breast, and flanks more or less distinctly flammulated with brown; under wing-coverts pale fulvous; bill brown; lower mandible whitish; feet blackish: whole length 4·7 inches, wing 2·5, tail 2·3. *Female* similar, but crest yellow or absent.

Hab. Veragua and Cisandean South America down to Argentine Republic.

This small Tyrant-bird is a summer visitor in the Plata district; it is shy and solitary; frequents woods and plantations, and perpetually utters, like the English Redstart, its sorrowful monotonous plaint, as it flits about in the upper foliage of the trees.

The nest is placed in a bush or low tree, and built of various soft materials compactly woven together, and the inside lined with feathers or vegetable down. The eggs are four, pale cream-colour, with large, well-defined spots of dark red.

The total length of this species is less than five inches. The prevailing colour of the plumage is rufous brown on the upper, and whitish brown on the under surface. Beneath the loose feathers of the crown there is a concealed spot of orange-red.

---

162. PYROCEPHALUS RUBINEUS (Bodd.).

(SCARLET TYRANT.)

Pyrocephalus rubineus, *Scl. et Salv. Nomencl.* p. 51; *Hudson, P. Z. S.* 1872, p. 808; *Durnford, Ibis,* 1877, p. 178 (Buenos Ayres); *Gibson, Ibis,* 1880, p. 27 (Buenos Ayres); *Barrows, Bull. Nutt. Orn. Cl.* vol. viii. p. 201 (Entrerios). **Pyrocephalus parvirostris,** *Burm. La-Plata Reise,* ii. p. 456 (Entrerios).

Description.—Above very dark cinereous, crested head and body below scarlet; bill and feet black: whole length 5·2 inches, wing 2·9, tail 2·3. *Female* above paler cinereous, below white; breast striated with cinereous; belly more or less rosy red.

Hab. S. America, from Colombia down to Buenos Ayres.

It is in vain, I think, to attempt to make more than one species out of this widely-spread bird, though specimens from the west coast are usually smaller.

The Scarlet Tyrant is about five inches and a half long; the neck, back, wings, and tail are black, all the rest of the plumage the most vivid scarlet imaginable. The loose feathers of the crown, which form a crest, are especially brilliant, and seem to glow like a live coal amidst the green foliage. Beside this bright Tyrant-bird even the Rainbow Tanagers look pale, and the "Jewel Humming-birds" decidedly sad-coloured. It is not strange, therefore, that in South America, where it has a very wide range, it is a species well known to the country people, and that they have bestowed on it many pretty names, most of which have reference to its splendid scarlet colour. In the Argentine Republic it is usually called *Churinche,* from its note, also *Federál* and *Fuegéro*; in other countries *Sangre de Toro* (bull's blood), and, better still, *Sangre*

Pura. Little Soldier and Coal of Fire are also amongst its names. The Guarani tribes call it *Guira-pitá* (red-bird); but another Indian name, mentioned by d'Orbigny, is the best—*Quarhi-rahi,* which signifies Sun-born.

The Churinche appears in Buenos Ayres about the end of September, and is usually first seen in localities to which Tyrant-birds are partial, such as low grassy grounds, with here and there a stalk or bush, and near a wood or plantation. Insects are most abundant in such places; and here the Churinche is seen, perched on a twig, darting at intervals to snap at the flies after the fashion of the Flycatchers, and frequently uttering its low, plaintive note. It is very common in the woods along the Plata; every orchard on the pampas is visited by a few of them; and they are very abundant about Buenos Ayres city. Going south they become rarer; but, strange to say, a few individuals find their way to the shores of the Rio Negro, though before reaching it they must cross a high, barren country quite unsuited to them. The natives of the Carmen have no name for the Churinche, but speak of it as a bird wonderful for its beauty and seldom seen. Amongst the dull-plumaged Patagonian species it certainly has a very brilliant appearance.

A very few days after their arrival the Churinches pair; and the male selects a spot for the nest—a fork in a tree from six to twelve feet from the ground, or sometimes a horizontal bough. This spot the male visits about once a minute, sits on it with his splendid crest elevated, tail spread out, and wings incessantly fluttering, while he pours out a continuous stream of silvery gurgling notes, so low they can scarcely be heard ten paces off, and somewhat resembling the sound of water running from a narrow-necked flask, but infinitely more rapid and musical. Of the little bird's homely, grey, silent mate the observer will scarcely obtain a glimpse, she appearing as yet to take little or no interest in the affairs that so much occupy the attention of her consort, and keep him in a state of such violent excitement. He is exceedingly pugnacious, so that when not fluttering on the site of his future nest, or snapping up some insect on the wing, he is eagerly pursuing other male Churinches, apparently bachelors, from tree to tree. At intervals he repeats his remarkable little song, composed of a succession of sweetly modulated metallic trills uttered on the wing. The bird usually mounts upwards from thirty to forty yards, and, with wings very much raised and rapidly vibrating, rises and drops almost perpendicularly half a yard's space five or six times, appearing to keep time to his notes in these motions. This song he frequently utters in the night, but without leaving his perch; and it then has a most pleasing effect, as it is less hurried, and the notes seem softer and more prolonged than when uttered by day. About a week after the birds have arrived, when the trees are only beginning to display their tender leaves, the nest is commenced. Strange to say, the female is the sole builder; for she now lays by her indifferent mien, and the art and industry she displays more than compensate for the absence of those beauties and accomplishments that make her mate so pleasing to the sight and ear. The materials of which the nest is composed are almost all gathered on trees; they are lichens, webs, and thistle-down: and the dexterity and rapidity with which they are gathered, the skill with which she disposes them, the tireless industry of the little bird, who visits her nest a hundred times an hour with invisible webs in her bill, are truly interesting to the observer. The lichens firmly held together with webs, and smoothly disposed with the tops outside, give to the nest the colour of the bark it is built on.

After the Churinche's nest is completed, the Bienteveo (*Pitangus bolivianus*) and the Common Cow-bird (*Molothrus bonariensis*) are the troublers of its peace. The first of these sometimes carries off the nest bodily to use it as material in building its own; the female Cow-bird is ever on the look out for a receptacle for her eggs. Seldom, however, does she succeed in gaining admittance to the Churinche's nest, as he is extremely vigilant and violent in repelling intruders. But his vigilance at times avails not; the subtle bird has watched and waited till, seizing a moment when the little Scarlet Tyrant is off his guard, she drops her surreptitious egg into his nest. When this happens, the Churinches immediately leave their nest. The nest is sometimes lined with feathers, but usually with thistle-down; the eggs are four, pointed, and spotted at the broad end with black; usually each egg has also a few large grey spots. The young are at first grey, marked with pale rufous, but soon become entirely grey, like the female. In about a month's time the belly of the males begins to assume a pale mauve-red; this spreads upwards towards the breast and throat; and finally the crest also takes on this colour. The Churinches raise two broods in a season—but if the nest is destroyed, will lay as many as four times.

The Scarlet Tyrant is the first of our summer visitors to leave us. As early as the end of January, and so soon as the young of the second brood are able to feed themselves, the adults disappear. Their going is not gradual, but they all vanish at once. The departure of all other migratory species takes place after a very sensible change in the temperature; but at the end of January the heat is unmitigated—it is, in fact, often greater than during December.

When the adults have gone, the silent young birds remain. Within a month's time the sexes of these may be distinguished. After another month the males begin to sing, and are frequently seen pursuing one another over the fields. It is only at the end of April, three months after the old birds have disappeared, that the young also take their departure. This is one of the strangest facts I have encountered in the migration of birds. The autumnal cold and wet weather seems to be the immediate cause of the young birds' departure; but in the adults, migration appears to be an instinct quite independent of atmospheric changes.

---

163. EMPIDONAX BIMACULATUS (Lafr. et d'Orb.).

(WING-BANDED TYRANT.)

Empidochanes argentinus, *Cab. J. f. O.* 1868, p. 196. **Empidonax brunneus,** *Ridgw. N. A. B.* ii. p. 363 (Paraná). **Empidonax bimaculatus,** *Scl. Ibis,* 1887, p. 65.

Description.—Above umber-brown, more or less rufescent; lores with a whitish spot; wings blackish, all the coverts broadly tipped with pale rufous, forming two transverse bars; outer margins of external secondaries of the same colour; tail brown, but not rufescent; beneath dirty cinereous white, throat and belly brighter, and with a yellowish tinge; under wing-coverts and inner margins of wing-feathers ochraceous; upper mandible dark brown, lower whitish; feet pale brown: whole length 5·0 inches, wing 2·6, tail 3·4.

Hab. S.E. Brazil, Bolivia, and N. Argentina.

This obscure species occurs in the northern wooded districts of Argentina.

---

164. CONTOPUS BRACHYRHYNCHUS, Cab.

(SHORT-BILLED TYRANT.)

Contopus brachyrhynchus, *Cab. J. f. O.* 1883, p. 214.

Description.—Above cinereous, lores whitish; wings and tail blackish, with slight whitish edgings to the wing-coverts and outer secondaries; beneath paler, whitish in the middle of the belly; flanks with a concealed white patch; bill above brown, beneath pale; feet black: whole length 7·0 inches, wing 4·0, tail 3·2.

Hab. Northern Argentina.

Herr Schulz, who discovered this species near Tucuman, tells us that it is a summer visitor, and is usually seen perched on the tops of the highest trees on the look-out for insects.

---

165. CONTOPUS BRACHYTARSUS, Scl.

(SHORT-FOOTED TYRANT.)

Contopus brachytarsus, *Scl. et Salv. Nomencl.* p. 52; *White, P. Z. S.* 1882, p. 608 (Salta).

Description.—Above dark plumbeous olive; crown darker, blackish; wings and tail blackish; the wing-coverts and outer secondaries more or less edged with whitish; beneath dirty white, clearer on the throat and middle of the belly, which latter has sometimes an olivaceous tinge; bill above blackish, beneath yellowish white; feet blackish; first primary shorter than the fifth: whole length 5·3 inches, wing 2·9, tail 2·5. *Female* similar.

Hab. Central and South America.

White found this widely ranging Tyrant "not uncommon in the forests of Salta."

---

166. MYIARCHUS TYRANNULUS (Müll.).

(RUSTY-TAILED TYRANT.)

Suiriri pardo y roxo, *Azara, Apunt.* ii. p. 143. **Myiarchus erythrocercus,** *Scl. et Salv. Nomencl.* p. 52.

Description.—Above brownish cinereous, crown rather darker; wings blackish, primaries narrowly edged with rufous, secondaries and coverts more broadly with dirty white; tail blackish, all the lateral rectrices with the greater part of the inner web rufous, leaving only a narrow blackish border alongside the shaft; beneath, throat and breast pale cinereous; belly and under wing-coverts pale sulphur-yellow; inner margin of rectrices pale rufous; bill dark horn-colour; feet blackish: whole length 7·4 inches, wing 3·8, tail 3·2. *Female* similar.

Hab. South America down to Argentina.

An example of this species, now in the British Museum, was procured by White in Catamarca.

---

167. MYIARCHUS FEROX (Gm.).

(FIERCE TYRANT.)

Myiarchus tyrannulus, *Scl. et Salv. Nomencl.* p. 52; *Durnford, Ibis,* 1878, p. 61 (Buenos Ayres); *White, P. Z. S.* 1882, p. 608 (Salta); *Barrows, Bull. Nutt. Orn. Cl.* vol. viii. p. 202 (Entrerios). **Myiarchus ferocior,** *Cab. J. f. O.* 1883, p. 214 (Tucuman).

Description.—Above dark cinereous, more or less olivaceous; wings and tail blackish; wing-coverts and outer secondaries with more or less defined edgings of dirty white; beneath, throat and breast cinereous, abdomen and under wing-coverts sulphur-yellow; bill dark brown; feet blackish: whole length 7·0 inches, wing 3·6, tail 3·4. *Female* similar.

Hab. Southern Antilles, and South America down to Argentine Republic.

There has been great confusion between this species and *M. tyrannulus,* from which the present bird may be distinguished by the absence of the rufous edgings to the inner webs of the rectrices.

Examples of *M. ferox* are in the British Museum from Punta Lara (*Durnford*), Mendoza (*Weisshaupt*), and Buenos Ayres (*Haslehust*).

---

168. MYIARCHUS ATRICEPS, Cab.

(BLACK-HEADED TYRANT.)

Myiarchus atriceps, *Cab. J. f. O.* 1883, p. 215.

Description.—Above greenish olive, cap black; wings and tail blackish, more or less margined with brownish; beneath, throat and neck pale grey; abdomen and under wing-coverts pale sulphur-yellow; inner margins of wing-feathers fulvous; bill dark horn-colour; feet black: whole length 7·0 inches, wing 3·7, tail 3·5. *Female* similar.

Hab. N. Argentina, Bolivia, and S. Peru.

Schulz found this species as a summer visitor in Tucuman.

---

169. EMPIDONOMUS AURANTIO-ATRO-CRISTATUS (d'Orb. et Lafr.).

(BLACK-AND-YELLOW-CRESTED TYRANT.)

Tyrannus aurantio-atro-cristatus, *d'Orb. Voy., Ois.* p. 312 (Corrientes); *Scl. et Salv. Nomencl.* p. 53; *Barrows, Bull. Nutt. Orn. Club,* vol. viii. p. 202 (Entrerios). **Tyrannus aurantio-atro-cristatus,** *Burm. La-Plata Reise,* ii. p. 453 (Rio Uruguay, Entrerios, Mendoza).

Description.—Above cinereous; cap shortly crested, black, with a large vertical spot of bright yellow; wings and tail brownish black, wing-coverts and secondaries slightly edged with whitish; beneath as above but rather paler, and with a very slight yellow tinge on the crissum; bill and feet black: whole length 6·5 inches, wing 3·8, tail 3·1. *Female* similar, but outer primaries less acuminated.

Hab. Interior of Brazil, Eastern Peru, Bolivia, and Argentina down to Buenos Ayres.

Alcide d'Orbigny met with this fine species in Corrientes, and Dr. Burmeister in Entrerios, and again near Mendoza. In the neighbourhood of Concepcion Mr. Barrows speaks of it as a "not very abundant summer resident, but one not easily overlooked, owing to its habit of perching on the topmost twig of any tree on which it alights, making forays from time to time, when tempted by its winged prey."

In the vicinity of Buenos Ayres likewise this Tyrant is not a common species. Like other birds of its genus it has an easy, rapid flight, and perches on trees or other elevated places, from which it occasionally makes a dash at passing insects. The nest, as in *T. melancholicus*, is a very slight structure of slender sticks, and the eggs are four, parchment colour, and spotted at the large end with dark brown or chocolate. Mr. Barrows found a Cowbird's egg in a nest of this species, which makes me think that it is less vigilant and warlike than *T. melancholicus*.

---

170. TYRANNUS MELANCHOLICUS, Vieill.

(MELANCHOLY TYRANT.)

Tyrannus melancholicus, *Burm. La-Plata Reise*, ii. p. 452; *Scl. et Salv. Nomencl.* p. 53; *Durnford, Ibis*, 1877, p. 178 (Buenos Ayres); *White, P. Z. S.* 1882, p. 608 (Salta); *Barrows, Bull. Nutt. Orn. Cl.* vol. viii. p. 202 (Entrerios).

Description.—Above grey with a slight greenish tinge; head with a concealed vertical crest of scarlet and yellow; lores and ear-coverts blackish; wings and tail brownish black with more or less of paler margins; beneath yellow, throat greyish white, breast more or less greyish, under wing-coverts pale yellow; bill and feet black; outer primaries attenuated; tail deeply forked: whole length 8·5 inches, wing 4·6, tail 4·0. *Female* similar.

Hab. Mexico and Central and South America down to Buenos Ayres.

The violent and bold temper exhibited by most Tyrant-birds during the breeding-season, a quality from which is derived the name of the family, is perhaps carried to a greater degree in this species than in any other; and when one spends many days or weeks in the marshy, littoral forests, where the bird is most abundant, and hears its incessant distressful screams, the specific name *melancholicus* does not seem altogether inappropriate; and that is the most that can be said of any specific name invented by science, and which does not merely describe some peculiarity of form or colour.

This Tyrant is one of the largest of its kind, its total length being nearly nine inches. The wings are long and suited for an aerial life; the legs are exceedingly short, and the feet are used for perching only, for this species never alights on the ground. The throat and upper parts are grey, tinged with olive on the back; the wings and tail dark; the breast yellow tinged with green; the belly pure yellow. Under the loose grey feathers of the crown is a fiery orange crest displayed in moments of excitement.

In Buenos Ayres these birds arrive in September, after which their shrill, angry cries are incessantly heard, while the birds are seen pursuing each other through the air or in and out amongst the trees—perpetually driven about by the contending passions of love, jealousy, and rage. As soon as their domestic broils are over, a fresh war against the whole feathered race begins, which does not cease until the business of propagation is finished. I have frequently spent hours watching the male, successively attacking, with scarcely an interval of rest, every bird, big or little, approaching the sacred tree where its nest was placed. Its indignation at the sight of a cowardly Carrion-Hawk (*Milvago*) skulking about in search of small birds' nests, and the boundless fury of its onset, were wonderful to witness.

They are extremely active, and when not engaged in their endless aerial battles, are pursuing large insects on the wing, usually returning after each capture to their stand, from which they keep a jealous watch on the movements of all winged things about them. They are fond of marshy places and water-courses, where they perch on a tall stalk to watch for insects, and also frequently skim over the water like Swallows to drink and dip their feathers.

A tall tree is usually selected for the nest, which is not unfrequently placed on the very topmost twigs, exposed to the sight of every creature passing overhead, and as if in defiance of birds of prey. With such an aggressive temper as this bird possesses it is not strange perhaps that it builds in the most exposed places, from which the female, in the absence of her vigilant consort, can keep a sharp eye on the movements of her feathered neighbours; but I have often thought it singular that they do not make a deeper receptacle for their eggs, for the nest is merely a slight platform of slender sticks, and very ill-adapted to retain its burden during high winds. The parasitical Cowbird never enters this nest, which is not strange.

The eggs are four in number, small for the bird, pointed, parchment-white, spotted with dark brown at the larger end.

---

171. MILVULUS TYRANNUS (Linn.).
(SCISSOR-TAIL TYRANT.)

Milvulus tyrannus, *Scl. et Salv. Nomencl.* p. 53; *Durnford, Ibis,* 1877, p. 178 (Buenos Ayres); *Gibson, Ibis,* 1880, p. 26 (Buenos Ayres); *Barrows, Bull. Nutt. Orn. Cl.* vol. viii. p. 203 (Entrerios). **Tyrannus violentus,** *Burm. La-Plata Reise,* ii. p. 453.

Description.—Above cinereous, rump blackish; cap jet-black, with a concealed yellow vertical crest; wings dark brown; tail black, outer web of the outer rectrix white; beneath white; bill and feet black; three outer primaries excised at the tips: whole length 14·0 inches, wing 4·6, tail 12·0. *Female* similar, but outer tail-feathers not so long.

Hab. Mexico, and Central and South America, down to Patagonia.

The *Tijereta* (Scissor-tail)—a name derived from the habit the bird has of opening and closing the two long feathers of the tail when flying—is found throughout South America, and in the summer of the Southern Hemisphere ranges as far south as Patagonia.

The tail is forked, and the two outer feathers exceed by over four inches in length the next two. The total length of the adult male is fourteen inches, the tail being ten inches long; this species is therefore one of the longest-tailed we know of. The tail of the female is about two inches shorter than that of the male. The head is intense black; the plumage of the crown is rather long and loose, and when raised displays a vivid yellow crest. The neck and upper surface is light, clear grey; the under surface pure white; the tail black. During flight the two long feathers of the tail stream out behind like a pair of black ribbons; frequently the bird pauses suddenly in its flight, and then the two long feathers open out in the form of the letter V.

The Scissor-tail is migratory, and arrives, already mated, at Buenos Ayres at the end of September, and takes its departure at the end of February in families—old and young birds together. In disposition and general habits it resembles the true Tyrant-birds, but differs from them in language, its various chirping and twittering notes having a hard percussive sound, which Azara well compares to the snapping of castanets. It prefers open situations with scattered trees and bushes; and is also partial to marshy grounds, where it takes up a position on an elevated stalk to watch for insects, and seizes them in the air like the

Flycatcher. It also greedily devours elderberries and other small fruits.

The nest is not deep, but is much more elaborately constructed than is usual with the Tyrants. Soft materials are preferred, and in many cases the nests are composed almost exclusively of wool. The inside is cup-shaped, with a flat bottom, and is smooth and hard, the thistle-down with which it is lined being cemented with gum. The eggs are four, sharply pointed, light cream-colour, and spotted, chiefly at the large end, with chocolate. In the breeding-time these Tyrants attack other birds approaching the nest with great spirit, and have a particular hatred to the Chimango, pursuing it with the greatest violence through the air with angry notes, resembling in sound the whetting of a scythe, but uttered with great rapidity and emphasis. How greatly this species is imposed upon by the Cow-bird, notwithstanding its pugnacious temper, we have already seen in my account of that bird.

The Scissor-tails have one remarkable habit; they are not gregarious, but once every day, just before the sun sets, all the birds living near together rise to the tops of the trees, calling to one another with loud, excited chirps, and then mount upwards like rockets to a great height in the air; then, after whirling about for a few moments, they precipitate themselves downwards with the greatest violence, opening and shutting their tails during their wild zigzag flight, and uttering a succession of sharp, grinding notes. After this curious performance they separate in pairs, and perching on the tree-tops each couple utters together its rattling castanet notes, after which the company breaks up.

Fam. XIV. PIPRIDÆ, or MANIKINS.

The brilliantly coloured *Pipridæ* or *Manikins* are nearly altogether confined to the tropical portions of the Neotropical Region, where they number about 70 species. Only one of these has as yet been discovered intruding in the northern outskirts of the Argentine Republic.

172. CHIROXIPHIA CAUDATA (Shaw).

(LONG-TAILED MANIKIN.)

Chiroxiphia caudata, *Scl. et Salv. Nomencl.* p. 55; *White, P. Z. S.* 1882, p. 608 (Misiones).

Description.—Above blue; cap scarlet; sides of head, nape, and wings black; tail black edged with bluish, two middle rectrices lengthened; beneath blue; throat, crissum, and under wing-coverts black; bill and feet reddish: whole length 6·0 inches, wing 3·1, tail 2·5. *Female* green, cap scarlet.

Hab. S.E. Brazil, Paraguay, and N.E. Argentina.

White obtained two or three males and one female of this Manikin in the forests of Misiones, on the banks of the Uruguay. One of his specimens is now in the British Museum.

Fam. XV. COTINGIDÆ, OR COTINGAS.

The *Cotingidæ* are another characteristic Neotropical family, mostly of splendid plumage, and nearly altogether confined within the limits of the tropics. Two stragglers only, belonging to the more obscure sections of the group, are as yet known to occur within the confines of Argentina, though it is quite probable that others may be found later on, when the northern forests are more completely explored.

---

173. PACHYRHAMPHUS POLYCHROPTERUS (Vieill.).

(WHITE-WINGED BÉCARD.)

Pachyrhamphus polychropterus, *Scl. et Salv. Nomencl.* p. 56; *Durnford, Ibis,* 1878, p. 61 (Buenos Ayres).

Description.—Above cinereous, upper back blackish; cap shining black; wings black, margins of coverts and secondaries white; tail black, four outer pairs of rectrices tipped with white; beneath cinereous, paler on the middle of the belly; under wing-coverts pale grey; bill and feet blackish: whole length 6·5 inches, wing 3·0, tail 2·6. *Female* above dull green, below yellowish; wings margined with rufous.

Hab. South Brazil, Paraguay, and N.E. Argentina.

This pretty little bird, the only species of the large South-American family *Cotingidæ* with which I am acquainted, comes as far south as Buenos Ayres, but is very scarce. It lives in woods, and is a shy, solitary bird, with nothing in its flight and general appearance to distinguish it from a Tyrant-bird. When flying, it utters a whistling note.

In January 1887, Durnford met with a pair of this species of Bécard in the riverain wood near Belgrano, and secured the male. They were busy catching flies, making frequent sallies from a willow tree in pursuit of them. Mr. Barrows obtained three specimens of what was probably the same bird at Concepcion in Entrerios, in November 1886 (*see* Bull. Nutt. Orn. Cl. viii. p. 203).

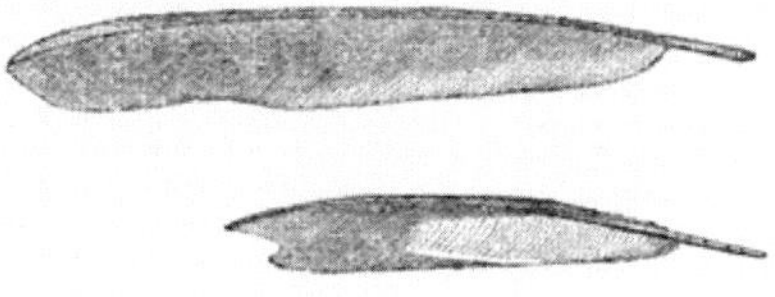

a. First primary of a male *Pachyrhamphus.*
b. Second ditto.

In the *male* of this species, as in many other Bécards, the second primary is abnormally shortened, being only about one inch in length. See remarks on this subject in P. Z. S. 1857, p. 72, whence the woodcut exhibiting this strange feature is taken by kind permission.

---

174. CASIORNIS RUBRA (Vieill.).
(RUFOUS CHEESE-BIRD.)

Suiriri roxo, *Azara, Apunt.* ii. p. 128. **Casiornis rubra**, *Scl. et Salv. Nomencl.* p. 57.

Description.—Above uniform ferruginous, lores paler; beneath lighter, belly yellowish; bill horn-colour, yellowish at the base; feet plumbeous: whole length 6·8 inches, wing 3·4, tail 3·1. *Female* similar.

Hab. S. Brazil, Paraguay, and N. Argentina.

White obtained a female specimen of this bird at Campo Colorado, Oran, in November 1880; it is now in Sclater's Collection.

Fam. XVI. PHYTOTOMIDÆ, OR PLANT-CUTTERS.

The peculiar form *Phytotoma,* remarkable for its toothed Fringilline bill, was associated by the older authors with the Finches. But modern researches have shown that it is not an Oscinine genus, and that its true place is near the *Cotingidæ*; indeed, some authors have placed it within the limits of that family.

Of the four known species of Plant-cutters, all restricted to South America, one is a well-known denizen of the Argentine Republic.

175. PHYTOTOMA RUTILA, Vieill.

(RED-BREASTED PLANT-CUTTER.)

[PLATE VIII.]

Phytotoma rutila, *Burm. La-Plata Reise,* ii. p. 451 (Paraná, Mendoza, Cordova, Tucuman, Catamarca); *Scl. et Salv. Nomencl.* p. 60; *Hudson, P. Z. S.* 1872, p. 537 (Rio Negro); *Barrows, Bull. Nutt. Orn. Cl.* viii. p. 203 (Entrerios); *White, P. Z. S.* 1882, p. 609 (Catamarca).

Description.—Above plumbeous, with slight darker shaft-spots; front of head bright red; wings and tail blackish, two well-marked wing-bars and tips of all lateral rectrices white; beneath bright red; flanks plumbeous; under wing-coverts whitish: whole length 7·0 inches, wing 3·5, tail 3·3. *Female*: above grey, densely striated with black; beneath dirty white, with dense black striations, belly and crissum fulvous.

Hab. Argentine Republic.

I found this curious little bird quite common in Patagonia, where the natives call it *Chingolo grande*, on account of its superficial resemblance to the common Song-Sparrow (*Zonotrichia pileata*). The colouring of the sexes differs considerably, the forehead and under surface of the male being deep brick-red; the upper parts dull grey, with a bar on the wing and the tips of the rectrices white; while in the female the upper parts are yellowish grey, obscurely mottled, and the breast and belly buff, with dark spots. In both sexes the eye is yellow, and the feathers of the crown pileated to form a crest.

This bird is usually seen singly, but sometimes associates in small flocks; it is resident, and a very weak flier, and feeds on tender buds and leaves, berries and small seed. The male is frequently seen perched on the summit of a bush, and, amidst the dull-plumaged species that people the grey thickets of Patagonia, the bright red bosom gives it almost a gay appearance. When singing, or uttering its alarm notes when the nest is approached, its voice resembles the feeble bleatings of a small kid or lamb. When approached it conceals itself in the bush, and when flying progresses by a series of short jerky undulations, the wings producing a loud humming sound.

The nest is made in the interior of a thorny bush, and built somewhat slightly of fine twigs and lined with fibres. The eggs are four, bluish-green in colour, with brownish flecks.

This species is found throughout the Argentine country, in dry, open situations, abounding with a scanty tree and bush vegetation.

PHYTOTOMA RUTILA ♂ et ♀

Suborder III. *TRACHEOPHONÆ.*

Fam. XVII. DENDROCOLAPTIDÆ, or WOOD-HEWERS.

The *Dendrocolaptidæ* are an important family in American Ornithology, numbering some 220 species, and distributed in greater or less abundance over every part of the Neotropical Region from Mexico to Patagonia. Within Argentine limits 46 species occur.

While green is the characteristic colour of the *Tyrannidæ,* brown is the favoured hue of the *Dendrocolaptidæ,* both the forest-loving and campos-frequenting members of the group being nearly without exception arrayed in various shades of that sombre colour, to which a ferruginous tail is a very frequent appendage.

The *Dendrocolaptidæ* fall into four subfamilies, all of which have representatives in Argentina. These are (1) the *Furnariinæ,* or Oven-birds, which are terrestrial in habits and have their feet adapted for this mode of life; (2) the *Sclerurinæ,* or Leaf-scrapers, known by their spiny tail, which keep to the ground inside the forests; (3) the *Synallaxinæ,* or Sharp-tails, mostly bush-frequenting birds; and (4) the *Dendrocolaptinæ,* or Wood-hewers, which have the habits of our Creepers (*Certhia*), and use their tail as a climbing-organ. All the members of this great family feed exclusively on insects.

Subfam. I. *FURNARIINÆ.*

176. GEOSITTA CUNICULARIA (Vieill.).

(COMMON MINER.)

Geositta cunicularia, *Burm. La-Plata Reise,* ii. p. 465 (Mendoza, Paraná); *Scl. et Salv. Nomencl.* p. 61; *Durnford, Ibis,* 1877, p. 178 (Buenos Ayres), et 1878, p. 395 (Chupat); *Barrows, Bull. Nutt. Orn. Cl.* viii. p. 203 (Entrerios).
Geositta tenuirostris, *White, P. Z. S.* 1882, p. 609 (Salta).

Description.—Above nearly uniform earthy brown; wing-feathers pale cinnamon-red; greater part of the outer webs, excepting the inner secondaries and a transverse bar across the secondaries, blackish; tail pale cinnamon-red, with a broad blackish band across the terminal half; beneath pale fulvous white, breast more or less variegated with blackish; under wing-coverts pale cinnamomeous; bill horn-colour, pale at the base; feet horn-colour: whole length 5·5 inches, wing 3·5, tail 2·0. *Female* similar.

Hab. Chili, Patagonia and Argentina.

The country people have a variety of names for this common and well-known species. In Buenos Ayres it is usually called *Manea-cola* (Shake-tail), in Patagonia *Caserita* (Little House-builder), and in other places *Minera* (Miner) or *Caminante* (Traveller), from its habit of running rapidly along a clean road or bridle-path before a person riding or walking.

It is a stout little bird, with very short toes quite unsuited for perching, and it does not, in fact, ever perch on a tree, though it manages to cling to a perpendicular bank very well, when engaged in opening its breeding-hole. It is resident and pairs for life, and lives in sterile places, feeding on small insects and spiders. In manner it is very lively, and runs swiftly over the bare ground, stopping very abruptly, then running on again, and at every pause slowly moving its half-open tail up and down. It flies swiftly, close to the ground, and always during its short flight trills out its clear, ringing, rapidly reiterated cry, which in sound resembles the laughter of a child.

On the grassy pampas the Miners invariably attach themselves to the *Vizcacheras*—as the groups of great burrows made by the large rodent, the Vizcacha, are called; for there is always a space free from grass surrounding the burrows where the birds can run freely about. In the sides of the deep pit-like entrance to one of these burrows the bird bores a cylindrical hole, from three to six feet long, and terminating in a circular chamber. This is lined with soft dry grass, and five white eggs are laid.

Though the birds inhabit the Vizcacha village all the year, they seem always to make a fresh hole to breed in every spring, the forsaken holes being given up to the small Swallow, *Atticora cyanoleuca*.

---

177. GEOBAMON RUFIPENNIS, Burm.

(RED-WINGED MINER.)

Geobamon rufipennis, *Burm. La-Plata Reise*, ii. p. 465 (Paraná).

Description.—Above reddish grey-brown; lores, rim round the eye, cheeks, and body below white; breast tinged with yellowish grey; wings blackish brown, inner webs ferruginous, with their tips and outer basal edges pale ferruginous; tail bright ferruginous, with a broad black transverse band near the tip; bill black, base of under mandible and legs pale brown: whole length 7·0 inches, wing 4·0, tail 2·0.

Hab. Paraná.

This form is unknown to us, and we can only give a short translation of Burmeister's description of it. It is said to resemble *Geositta*, but has a much shorter and perfectly straight beak.

---

178. FURNARIUS RUFUS (Gm.).

(RED OVEN-BIRD.)

Furnarius rufus, *Burm. La-Plata Reise*, ii. p. 462 (La Plata); *Scl. et Salv. Nomencl.* p. 61; *Durnford, Ibis*, 1877, p. 179 (Buenos Ayres); *Gibson, Ibis*, 1880, p. 16 (Buenos Ayres).

Description.—Above earthy brown, with a slight rufescent tinge, wing-feathers blackish, margined with pale brown; whole of the outer secondaries pale brown, like the back; tail and upper tail-coverts bright ferruginous brown; below white, breast and flanks and under wing-coverts pale sandy-brown; under surface of the wing with a broad sandy bar across the basal portion; bill and feet horn-colour: whole length 7·8 inches, wing 4·0, tail 2·8. *Female* similar.

Hab. Argentina, Uruguay, and Paraguay.

The Red Oven-bird is an extremely well-known species in Argentina, and, where found, a great favourite on account of its familiarity with man, its loud, ringing, cheerful voice, and its wonderful mud nest, which it prefers to build near a human habitation, often on a cornice, a projecting beam, or on the roof of the house itself.

It is a stout little bird, about 8½ to 9 inches long, with a slender, slightly-curved beak nearly an inch in length, and strong legs suited to its terrestrial habits. The upper plumage is uniform rufous-brown in colour, brightest on the tail; the under surface very light brown. It ranges throughout the Argentine Republic to Bahia Blanca in the south, and is usually named *Hornero* or *Casera* (Oven-bird or House-builder), but in Paraguay and Corrientes it is called *Alonzo Garsia* or *Alonzito.* Azara could give no reason for such a name; but it seems to me that one need not look for one beyond the fact that this species inspires an affectionate admiration in the country people: I mean in those of *Spanish* origin, for the bird-killing French and Italians have no tenderness for it. I have frequently been assured by natives that the Hornero is a *pious* bird, and always suspends its labours on sacred days. With this pretty belief about it in their minds, it is not strange that in some districts they have called it by a human name.

It is resident, pairs for life, and finds its food, which consists of larvæ and worms, exclusively on the ground. It delights in open places, where it can move freely about on the ground; and is partial to court-yards, clean garden-walks, &c., where, with head thrown back and bosom prominent, it struts along with an air of great gravity, lifting its foot high at each step, and holding it suspended for a moment in the air before setting it firmly down. I once saw one fly down on to a narrow plank about ten feet long lying out on the wet grass; it walked gravely to the end of the plank, then turned, and deliberately walked back to the other end, and so on for about twenty times, appearing to take the greatest pleasure in the mere act of promenading on a smooth level surface. When disturbed, the Oven-bird has a loud, monotonous note of alarm or curiosity, which never fails to bring all its fellows within hearing-distance to the spot. The movements of a fox, weasel, or cat in a plantation can always be known from the noisy turmoil among the Oven-birds. At frequent intervals during the day the male and female meet and express their joy in clear, resonant notes sung in concert—a habit common to a very large number of Dendrocolaptine birds, including, I think, all those species which pair for life. In a majority of species this vocal performance merely consists of a succession of confused notes or cries, uttered with great spirit and emphasis; in the Oven-bird it has developed into a kind of harmonious singing. Thus, the first bird, on the appearance of its mate flying to the place of meeting, emits loud measured notes, sometimes a continuous trilling note with a somewhat hollow metallic sound; but immediately on the other bird joining, this introductory passage is changed to rapid triplets, strongly accented on the first and last notes, while the second bird utters a series of loud measured notes perfectly according with the triplets of the first. While thus singing they stand facing each other, their necks outstretched, wings hanging, and tails spread, the first bird trembling with its rapid utterances, the second beating on the branch with its wings. The finale consists of three or four notes uttered by the second bird alone, and becoming successively louder and more piercing until the end. There is an infinite variety in the tone in which different couples sing, also in the order in which the different notes are uttered, and even the same couple do not repeat their duet in precisely the same way; but it is always a rhythmical and, to some extent, an harmonious performance, and as the voices have a ringing, joyous character, it always produces a pleasing effect on the mind.

In favourable seasons the Oven-birds begin building in the autumn, and the work is resumed during the winter whenever there is a spell of mild wet weather. Some of their structures are finished early in winter, others not until spring, everything depending on the weather and the condition of the birds. In cold dry weather, and when food is scarce, they do not work at all. The site chosen is a stout horizontal branch, or the top of a post, and they also frequently build on a cornice or the roof of a house; and sometimes, but rarely, on the ground. The material used is mud, with the addition of horsehair or slender fibrous rootlets, which make the structure harder and prevent it from cracking. I have frequently seen a bird, engaged in building, first pick up a thread or hair, then repair to a puddle, where it was worked into a pellet of mud about the size of a filbert, then carried to the nest. When finished the structure is shaped outwardly like a baker's oven, only with a deeper and narrower entrance. It is always placed very conspicuously, and with the entrance facing a building, if one be near, or if at a roadside it looks toward the road; the reason for this being, no doubt, that the bird keeps a cautious eye on the movements of people near it while building, and so leaves the nest opened and unfinished on that side until the last, and there the entrance is necessarily formed. When the structure has assumed the globular form with only a narrow opening, the wall on one side is curved inwards, reaching from the floor to the dome, and at the inner extremity an aperture is left to admit the bird to the interior or second chamber, in which the eggs are laid. A man's hand fits easily into the first or entrance chamber, but cannot be twisted about so as to reach the eggs in the interior cavity, the entrance being so small and high up. The interior is lined with dry soft grass, and five white pear-shaped eggs are laid. The *oven* is a foot or more in diameter, and is sometimes very massive, weighing eight or nine pounds, and so strong that, unless loosened by the swaying of the branch, it often remains unharmed for two or three years. The birds incubate by turns, and when one returns from the feeding-ground it sings its loud notes, on which the sitting bird rushes forth to join in the joyous chorus, and then flies away, the other taking its place on the eggs. The young are exceedingly garrulous, and when only half-fledged may be heard practising trills and duets in their secure oven, in shrill tremulous voices, which change to the usual hunger-cry of young birds when the parent enters with food. After leaving the nest, the old and young birds live for two or three months together, only one brood being raised in each year. A new oven is built every year, and I have more than once seen a second oven built on the top of the first, when this has been placed very advantageously, as on a projection and against a wall.

A very curious thing occurred at the estancia house of a neighbour of mine in Buenos Ayres one spring. A pair of Oven-birds built their oven on a beam-end projecting from the wall of a rancho. One morning one of the birds was found caught in a steel trap placed the evening before for rats, and both of its legs were crushed above the knee. On being liberated it flew up to and entered the oven, where it bled to death, no doubt, for it did not come out again. Its mate remained two days, calling incessantly, but there were no other birds of its kind in the place, and it eventually disappeared. Three days later it returned with a new mate, and immediately the two birds began carrying pellets of mud to the oven, with which they plastered up the entrance. Afterwards they built a second oven, using the sepulchre of the dead bird for its foundation, and here they reared their young. My neighbour, an old native, had watched the birds from the time the first oven was begun, feeling greatly interested in their diligent ways, and thinking their presence at his house a good omen; and it was not strange that, after witnessing the entombment of one that died, he was more convinced than ever that the little House-builders are "pious birds."

---

179. FURNARIUS TRICOLOR, Cab.

(CRESTED OVEN-BIRD.)

Furnarius tricolor, *Cab. Journ. f. Orn.* 1878, p. 196 (Cordova). **Furnarius figulus,** *White, P. Z. S.* 1882, p. 609 (?).

Description.—Above nearly uniform earthy brown, head slightly crested; tail bright ferruginous; beneath paler, throat pure white; middle of belly and crissum white; under surface of wings blackish, with a transverse bar of pale cinnamon; bill hazel, paler at the base; feet blackish: whole length 5·7 inches, wing 2·8, tail 2·2.

Hab. Cordova.

This pretty little species is a recent discovery of Döring in the Sierras of Cordova; it may be at once recognized by its small size and crested head.

---

180. UPUCERTHIA DUMETORIA (Geoffr. et d'Orb.).

(PATAGONIAN EARTH-CREEPER.)

Ochetorhynchus dumetorius, *Burm. La-Plata Reise,* ii. p. 463 (Mendoza). **Upucerthia dumetoria,** *Scl. et Salv. Nomencl.* p. 62; *Hudson, P. Z. S.* 1872, p. 544 (Rio Negro); *Durnford, Ibis,* 1877, p. 35, et 1878, p. 395 (Chupat); *White, P. Z. S.* 1883, p. 433 (Cordova).

Description.—Above earthy brown; long superciliary stripe pale ochraceous; wings blackish, with a broad transverse cinnamomeous bar; tail blackish, lateral rectrices tipped with pale cinnamon; beneath dirty white, clear white on the throat and middle of the belly; breast-feathers margined with blackish; under wing-coverts pale cinnamomeous; bill dark horn-colour, pale at the base; feet horn-colour: whole length 9·0 inches, wing 4·0, tail 3·5. *Female* similar.

Hab. Patagonia and Chili.

These birds are very common in Patagonia, being resident there; some individuals, however, migrate north in winter, and I once obtained a pair, male and female, near Buenos Ayres city in the month of June.

Their legs are short, but on the ground their movements are very rapid, and, like the Miner (*Geositta*) already described, they fly reluctantly, preferring to run rapidly from a person walking or riding, and at such times they look curiously like a very small Curlew with an extravagantly long beak. They are active, lively birds, and live in pairs, sometimes uniting in small, loose flocks; they are partial to places where scattered bushes grow on a dry sterile soil, and have a swift low flight; when flying they frequently utter a shrill, trilling, or rapidly reiterated note, in sound resembling laughter. In manners, flight, language, and colouring this bird closely resembles the smaller short-beaked *Geositta cunicularia*, and like that species it also breeds in deep holes in banks; but I am not able to say whether it excavates the breeding-hole or takes possession of one already made. Durnford found it breeding in a hole four feet deep in the bank of a dry lagoon. The nest was of dry grass and lined with the fur of the cavy. It contained three white eggs.

---

181. UPUCERTHIA RUFICAUDA (Meyen).

(RED-TAILED EARTH-CREEPER.)

Ochetorhynchus ruficauda, *Burm. La-Plata Reise*, ii. p. 463 (Mendoza).

Description.—Above earthy brown, superciliaries whitish, lower half of back and outer secondaries strongly tinged with rufous; tail deep ferruginous red, inner webs of all the lateral rectrices black; beneath white, breast more or less freckled with greyish; belly, flanks, and crissum pale cinnamomeous brown; under surface of wings blackish, with a transverse cinnamomeous bar; bill and feet blackish: whole length 8·0 inches, wing 3·5, tail 3·3. *Female* similar.

Hab. Chili and Mendoza.

The straight bill and red colour of the tail-feathers at once distinguish this species from the former. Burmeister obtained specimens of it in the Sierra of Uspallata, where it was met with hopping about the rocks and feeding on insects.

---

182. UPUCERTHIA LUSCINIA (Burm.).

(WARBLING EARTH-CREEPER.)

Ochetorhynchus luscinia, *Burm. La-Plata Reise*, ii. p. 464 (Mendoza, Paraná).
Upucerthia luscinia, *Scl. et Salv. Nomencl.* p. 62; *White, P. Z. S.* 1882, p. 610 (Catamarca); *Salvin, Ibis*, 1880, p. 358 (Salta).

Description.—Above earthy brown; front, lores, and rim round the eye bright rufous; wings blackish, outer webs more or less edged with rufous earthy brown; tail earthy brown, lateral rectrices tinged with rufous; below pale cinereous with a slight rufescent tinge; throat clear white; under wing-coverts and inner margins of the wing-feathers cinnamon-red; bill hazel, paler at the base; feet pale brown: whole length 7·5 inches, wing 3·0, tail 3·1. *Female* similar.

Hab. Argentina.

Professor Burmeister was the first discoverer of this species, which he tells us is common near Mendoza, in Paraná, and in the neighbouring pampas. In Paraná he found it nesting under the roof of his house and feeding upon insects. The eggs are pure white. It is the *Ruisiñor* or "Nightingale" of the natives, whence he gave it the specific name *luscinia*—a strange name for any species in the shrill-voiced Dendrocolaptine family.

---

183. CINCLODES FUSCUS (Vieill.).
(BROWN CINCLODES.)

Cinclodes fuscus, *Scl. et Salv. Nomencl.* p. 62; *Durnford, Ibis,* 1877, p. 179 (Buenos Ayres); *White, P. Z. S.* 1882, p. 610 (Catamarca); *Barrows, Bull. Nutt. Orn. Cl.* viii. p. 205 (Entrerios and Pampas). **Cinclodes vulgaris,** *Burm. La-Plata Reise,* ii. p. 463 (Mendoza, Paraná).

Description.—Above dark earthy brown, lores and superciliaries whitish; wings blackish, with a broad transverse cinnamomeous bar; outer tail-feathers blackish, broadly tipped with pale cinnamomeous white; beneath pale cinereous, with a cinnamomeous tinge; throat white, slightly spotted with blackish; bill and feet horn-colour: whole length 7·3 inches, wing 4·0, tail 3·0. *Female* similar.

Hab. Argentina, Chili, Peru, Bolivia, and Ecuador.

This homely little species differs considerably from most Dendrocolaptine birds in colour and habits; and being of a uniform dull, fuscous hue, its appearance is most uninteresting. It inhabits the whole of Patagonia, but is migratory, possessing, what is rare in this family, a powerful flight. In winter it is common all over the pampas and the Plata district, ranging north to Paraguay. It is always found near water, its favourite hunting-ground being the borders of a stream. On the ground its motions are quick and lively, but when perching on a tree it sits motionless in one position, and when attempting to move appears to lose its balance. These birds cannot be called strictly gregarious, but where abundant they are fond of gathering in loose flocks, sometimes numbering one or two hundred individuals, and when thus associating are very playful, frequently pursuing and wheeling about each other, and uttering a sharp, trilling note. On a warm day in winter they are occasionally heard attempting to sing, the bird darting up vertically into the air and pouring out with great energy a confused torrent of unmelodious sounds.

Their habits, so much less sedate and strikingly in contrast with those of most of the birds in this family, are no doubt due to the greater powers of flight possessed by *Cinclodes*.

---

184. CINCLODES BIFASCIATUS, Sclater.

(WHITE-WINGED CINCLODES.)

Cinclodes bifasciatus, *Sclater, P. Z. S.* 1858, p. 448.

Description.—Above earthy brown with a rufescent tinge on the back, superciliaries white; wings blackish, with a white bar on the secondaries and a second white bar on the inner primaries; tail blackish, outer rectrices tipped with white; beneath white, passing into greyish cinnamomeous on the flanks and crissum; bend of the wing and under wing-coverts white; bill and feet horn-colour: whole length 8·0 inches, wing 4·0, tail 3·3.

Hab. Bolivia and Eastern Argentina.

Weisshaupt obtained examples of this species in the vicinity of Mendoza, from one of which our description is taken.

---

185. HENICORNIS PHŒNICURUS (Gould).

(DARK-TAILED HENICORNIS.)

Henicornis phœnicurus, *Scl. et Salv. Nomencl.* p. 62; *Durnford, Ibis,* 1878, p. 395 (Central Patagonia).

Description.—Above earthy brown, slight superciliary line white, edges of outer secondaries, lower back, and upper tail-coverts bright ferruginous red; tail black, outer webs of outer tail-feathers and slight external edging of the others bright ferruginous; beneath, throat and breast pure white, belly cinereous, flanks tinged with rufous; crissum bright ferruginous; bill dark horn-colour, base of lower mandible yellowish; feet horn-colour: whole length 7·5 inches, wing 3·3, tail 3·0.

Hab. Chili and Patagonia.

Durnford met with this peculiar form in Central Patagonia in 1877–78. He says that it was resident and common among the bushes throughout his journey.

186. LOCHMIAS NEMATURA (Licht.).

(BRAZILIAN LOCHMIAS.)

Lochmias nematura, *Burm. La-Plata Reise,* ii. p. 462 (Paraná).

Description.—Above clear brown, rump blackish, elongated superciliaries white; tail blackish; beneath blackish brown, with numerous large white oval spots occupying the centre of the feathers; bill horn-colour, lower mandible pale brown at the base; feet brown: whole length 5·5 inches, wing 2·8, tail 2·0.

Hab. South-east Brazil and Northern Argentina.

Professor Burmeister met with this Brazilian species near Paraná, where it lives on the ground among the bushes.

Subfam. II. *SCLERURINÆ.*

187. SCLERURUS UMBRETTA (Licht.).

(THE SPINY LEAF-SCRAPER.)

Sclerurus caudacutus, *White, P. Z. S.* 1882, p. 610 (Misiones). **Sclerurus umbretta,** *Scl. et Salv. Nomencl.* p. 62.

Description.—Above dark brown, rump and upper tail-coverts rufous; wing-feathers blackish, glossed with dull brown; tail black; beneath, throat white, the feathers edged with dark brown; breast dull rufous; belly and flanks same colour as back; under wing-coverts whitish brown; bill and feet black: whole length 7·1 inches, wing 3·6, tail 2·6.

Hab. Brazil, Paraguay, and Argentina.

White has the following note on this interesting species:—"Only two of these birds were observed during my trip through Misiones. It frequents the dense gloomy forests, where it busies itself in scraping amongst the dead leaves; and although it may be close at hand and the rustling distinct, a quick eye is required to detect it, as its plumage is of the exact colour of decaying foliage. If startled, it flies up onto the trunk of the nearest tree, and there remains perfectly motionless in an upright position. I never heard it utter a single note."

Subfam. III. *SYNALLAXINÆ.*

188. PHLŒOCRYPTES MELANOPS (Vieill.).

(RUSH-LOVING SPINE-TAIL.)

Synallaxis melanops, *Burm. La-Plata Reise,* ii. p. 470 (Mendoza). **Phlœocryptes melanops,** *Scl. et Salv. Nomencl.* p. 63; *Durnford, Ibis,* 1877, p. 179 (Buenos Ayres), et 1878, p. 396 (Central Patagonia); *Barrows, Bull. Nutt. Orn. Cl.* viii. p. 206 (Bahia Blanca).

Description.—Above, forehead brown, crown blackish, broad superciliaries buffy white; upper half of back black, marked with a few grey stripes; lower back and rump, also sides of head and neck, light brown; wings blackish, mottled with light chestnut on the coverts; and a broad band of the same colour occupying the basal half of the wing-feathers; tail blackish, the two middle feathers brownish grey, the others slightly tipped with the same colour; beneath white, more or less tinged on the throat, flanks, and under tail-coverts with pale brown; under wing-coverts fulvous; bill and feet pale horn-colour: whole length 5·8 inches, wing 2·3, tail 1·6. *Female* similar.

Hab. Chili, Patagonia, and Argentina.

This is one of our few strictly migratory species in the family *Dendrocolaptidæ*. Probably it winters in South Brazil, as in the northern parts of the Argentine country it is said to be a summer visitor. On the pampas it appears in September, and all at once becomes very abundant in the rush-beds growing in the water, where alone it is found. The migration no doubt is very extensive, for in spring I found it abundant in the rush-beds in the Rio Negro valley, and Durnford met with it much further south on the river Sanguelen, a tributary of the Chupat. Migratory birds are, as a rule, very little given to wandering; that is to say, they do not go much beyond the limits of the little coppice, reed-bed, or spot of ground which they make their summer home, and this species is no exception. It spends the warm season secluded in its rush-bed: and when disturbed flies with great reluctance, fluttering feebly away to a distance of a few yards, and then dropping into the rushes again, apparently quite incapable of a sustained flight. How a bird so feeble on the wing, and retiring in its habits, is able to perform a long, annual migration, when in traversing vast tracts of open country it must be in great peril from rapacious kinds, is a great mystery. No doubt many perish while travelling; but there is this circumstance in their favour: an incredible number of birds of various kinds, many as weak and exposed to attack as the *Phlœocryptes*, migrate simultaneously; Hawks are very thinly scattered along their route, and as a rule these birds feed only once or twice a day, if the meals are large enough to fill the stomach, so that while the Hawk is inactive, digesting his meal, thousands of migrants have sped by on their journey and are beyond his reach for ever.

This Spine-tail seldom ventures out of its rush-bed, but is occasionally seen feeding in the grass and herbage a few yards removed from the water. Its language is peculiar, this being a long cicada-like note, followed by a series of sounds like smart taps on a piece of dry wood. It frequents the same places as the small Many-coloured Tyrant (*Cyanotis azaræ*), and these little neighbours, being equally inquisitive, whenever a person approaches the rushes often emerge together, one uttering wooden-sounding creaks and raps, the other liquid gurgling notes—a little brown bird and a little bird with many bright colours, both, in very different tones, demanding to know the reason of the intrusion.

The nest is a very wonderful structure, and is usually attached to three upright stems; it is domed, oval-shaped, about nine inches deep, and the small circular aperture which is close to the top is protected by a sloping tile-like projection. It is built of tough grass-leaves, which are apparently first daubed with wet clay and then ingeniously woven in, with the addition, I think, of some kind of mucilage: the whole nest is, when finished, light but very strong, and impervious to wet. Until the rushes die and drop the nest remains securely fastened to them, and in winter affords a safe and comfortable retreat to the small reed-frogs, of which sometimes as many as three or four are found living in one nest. The interior is very thickly lined with feathers; the eggs are three, pear-shaped, and a bright, beautiful blue colour, sometimes with a slight greenish tinge.

The bird is so abundant in extensive marshes that I have on several occasions, during a day's ramble, found as many as forty or fifty nests, sometimes a dozen or more being placed close together, but I have never taken more than three eggs from one nest. I mention this because I have seen it stated that four or five eggs are sometimes found.

I trust that no reader of this sketch imagines that I robbed all the eggs contained in so many nests. I did nothing so barbarous, although it is perhaps "prattling out of fashion" to say so; but with the destructive, useless egg-collecting passion I have no sympathy. By bending the pliant rushes downwards the eggs can be made to roll out into the hand; and all those which I thus took out to count were, I am glad to say, put back in their wonderful cradles. I had a special object in examining so many nests. A gaucho boy once brought me a nest which had a small circular *stopper*, made of the same texture as the body of the nest, attached to the aperture at the *side*, and when swung round into it fitting it as perfectly as the lid of the trap-door spider fits the burrow. I have no doubt that it was used to close the nest when the bird was away, perhaps to prevent the intrusion of reed-frogs or of other small birds; but I have never found another nest like it, nor have I heard of one being found by any one else; and that one nest, with its perfectly-fitting stopper, has been a puzzle to my mind ever since I saw it.

189. LEPTASTHENURA ÆGITHALOIDES (Kittl.).

(TIT-LIKE SPINE-TAIL.)

Synallaxis ægithaloides, *Burm. La-Plata Reise,* ii. p. 469 (Mendoza); *Hudson, P. Z. S.* 1872, p. 544 (Rio Negro). **Leptasthenura ægithaloides,** *Scl. et Salv. P. Z. S.* 1869, p. 632 (Buenos Ayres); *iid. Nomencl.* p. 63; *Durnford, Ibis,* 1877, p. 180 (Buenos Ayres), et 1878, p. 396 (Centr. Patagonia); *Gibson, Ibis,* 1880, p. 30 (Buenos Ayres); *White, P. Z. S.* 1882, p. 611 (Catamarca); *Barrows, Bull. Nutt. Orn. Cl.* viii. p. 206 (Entrerios).

Description.—Above pale earthy brown; crown black, striped with clear brown; lores, sides of head, and throat white, with minute black spots; wings blackish, the edges of the outer webs of the primaries and the basal part of the secondaries light rufescent brown; tail black, lateral rectrices tipped and margined with pale grey; beneath pale grey, throat white; bill and feet horn-colour: whole length 6·2 inches, wing 2·4, tail 3·5. *Female* similar.

Hab. Chili, Argentina, and Patagonia.

This is a restless little bird, seen singly or in parties of three or four. In manner and appearance it resembles the Long-tailed Titmouse (*Parus*), as it diligently searches for small insects in the trees and bushes, frequently hanging head downwards to explore the under surface of a leaf or twig, and while thus engaged continually uttering a little sharp querulous note. They are not migratory, but in winter seem to wander about from place to place a great deal; and in Patagonia, in the cold season, I have frequently seen them uniting in flocks of thirty or forty individuals, and associating with numbers of Spine-tails of other species, chiefly with *Synallaxis sordida,* and all together advancing through the thicket, carefully exploring every bush in their way.

D'Orbigny says that it makes a nest of rootlets and moss in a bush; but where I have observed this bird it invariably breeds in a hole in a tree, or in the nest of some other bird, often in the clay structure of the Oven-bird. But in Patagonia, where the Oven-bird is not known, this Spine-tail almost always selects the nest of the *Synallaxis sordida.* It carries in a great deal of soft material—soft grass, wool, and feathers—to reline the cavity, and lays five or six, white, pointed eggs.

---

190. LEPTASTHENURA FULIGINICEPS (d'Orb. et Lafr.).

(BROWN-CRESTED SPINE-TAIL.)

Leptasthenura fuliginiceps, *Scl. et Salv. Nomencl.* p. 63; *White, P. Z. S.* 1882, p. 611 (Catamarca). **Synallaxis fuliginiceps,** *Burm. La-Plata Reise,* ii. p. 469.

Description.—Above pale earthy brown; forehead and slightly crested crown rufous brown; wings blackish, edges of outer webs of all the wing-feathers and basal part of the secondaries chestnut; tail clear brown; beneath paler, earthy brown; bill and feet horn-colour: whole length 6·0 inches, wing 2·2, tail 3·1.

Hab. Bolivia and N. Argentina.

This species, discovered by d'Orbigny in Bolivia, was met with by White in Catamarca, "on the slopes of the hills, outside the dense wood," and by Prof. Burmeister near Paraná.

---

191. SYNALLAXIS FRONTALIS, Pelz.

(BROWN-FRONTED SPINE-TAIL.)

Synallaxis frontalis, *Scl. et Salv. Nomencl.* p. 63; *Scl. P. Z. S.* 1874, p. 8; *Salvin, Ibis,* 1880, p. 358 (Salta); *White, P. Z. S.* 1882, p. 611 (Salta, Catamarca).
Synallaxis ruficapilla, *d'Orb. Voy., Ois.* p. 246 (Corrientes); *Burm. La-Plata Reise,* ii. p. 468 (Paraná).

Description.—Above earthy brown; cap chestnut; front earthy brown; wing-coverts chestnut, wing-feathers olive-brown, the outer webs edged with chestnut; tail chestnut; beneath, throat blackish, with slight whitish mottlings; breast, sides, and under tail-coverts pale earthy brown, belly brownish white; under wing-coverts fulvous; bill and feet horn-colour: whole length 5·6 inches, wing 2·2, tail 3·0.

Hab. South America, from Colombia to Argentina.

---

192. SYNALLAXIS SUPERCILIOSA, Cab.

(EYEBROWED SPINE-TAIL.)

Synallaxis superciliosa, *Cab. J. f. O.* 1883, p. 110 (Tucuman).

Description.—Above, head on top bright chestnut, lores white, superciliaries yellowish white; sides of head, neck, and back earthy brown inclining to olive; upper wing-coverts chestnut, wing-feathers blackish, the webs of the outer margins dull chestnut; tail chestnut; beneath, upper half of throat white, lower half black tipped with white; breast and belly brownish white; flanks and under tail-coverts pale earthy brown; under wing-coverts fulvous; bill and feet horn-colour: whole length 7·0 inches, wing 2·0, tail 3·5.

Hab. Tucuman.

This Spine-tail, so far as we know at present, is peculiar to Tucuman, where it was discovered by Herr Schulz.

---

193. SYNALLAXIS SPIXI, Scl.

(SPIX'S SPINE-TAIL.)

Synallaxis spixi, *Burm. P. Z. S.* 1868, p. 636 (Buenos Ayres); *Scl. et Salv. P. Z. S.* 1869, p. 632 (Buenos Ayres); *Scl. P. Z. S.* 1874, p. 9.

Description.—Above, crown chestnut; lores and sides of head dark cinereous; hind neck, back, also wing- and tail-feathers olive-brown; upper wing-coverts chestnut; beneath dark cinereous, becoming whitish on the belly, throat blackish; under wing-coverts fulvous chestnut; bill black, feet horn-colour; whole length 6·7 inches, wing 2·0, tail 3·2.

Hab. Southern Brazil and Argentina.

I like Azara's name *Chicli,* which, to one acquainted with the habits of this and of the following species, seems very appropriate, suggesting, as I imagine it does, a small creature possessing a sharp two-syllabled note; for although Hartlaub, in his Nomenclature of Azara, gives *S. ruficapilla* as the species meant by *Chicli,* the account of its habits in the 'Apuntamientos' seems to point to *S. spixi* or to *S. albescens.*

Azara says:—"I give it this name because it sings it plainly, in a loud sharp tone, which may be heard at a distance, repeating it so frequently that the pauses last no longer than the sound. It is resident (in Paraguay), solitary and not abundant: inhabits thickets of aloes and thorn, without rising more than two yards above the surface, or showing itself in open places. It moves about incessantly, but does not leave its thicket to visit the woods or open ground, its flight being only from bush to bush; and though it is not timid, it is hard to detect it in its stronghold, and to hear it one would imagine that it was perched overhead on a tree, when it is hidden all the time in the brushwood at the roots."

This habit of concealing itself so closely inclines me to think that this species, rather than *S. albescens,* was the bird described by Azara, although in both species the language is nearly the same. I have nothing to add to the above account from the 'Apuntamientos,' except that in the love-season this species has a low, strange-sounding little song, utterly unlike its usual strident cry. When singing, it sits motionless on the summit of a low bush in a dejected attitude with head drawn in, and murmurs its mysterious little melody at intervals of half a minute.

---

194. SYNALLAXIS ALBESCENS, Temm.

(WHITE-THROATED SPINE-TAIL.)

Synallaxis albescens, *Scl. et Salv. Nomencl.* p. 63; *Scl. P. Z. S.* 1874, p. 9; *Durnford, Ibis,* 1877, p. 180 (Buenos Ayres); *White, P. Z. S.* 1882, p. 611 (Misiones); *Barrows, Bull. Nutt. Orn. Cl.* viii. p. 207 (Entrerios).

Description.—Above, forehead grey, crown pale chestnut; sides of head and neck, back, and tail pale earthy brown; upper wing-coverts pale chestnut, wing-feathers olive-brown; beneath white, faintly washed with earthy brown; under wing-coverts fulvous: whole length 5·3 inches, wing 2·0, tail 2·2.

Hab. S. America, from Veragua to Buenos Ayres.

This species, although by no means abundant in Buenos Ayres, is met with much more frequently than the *S. spixi*, which it closely resembles in size, colour, habits, and language. It is, indeed, an unusual thing for two species so closely allied to be found inhabiting the same district. In both birds the colours are arranged in precisely the same way; but the chestnut tint on *S. albescens* is not nearly so deep, the browns and greys are paler, and there is less black on the throat.

I am pretty sure that in Buenos Ayres it is migratory, and as soon as it appears in spring it announces its arrival by its harsh, persistent, two-syllabled note, wonderfully strong for so small a bird, and which it repeats at intervals of two or three seconds for half an hour without intermission. When close at hand it is quite as distressing as the grinding noise of a Cicada. This painful noise is uttered while the bird sits concealed amid the foliage of a tree, and is renewed at frequent intervals, and continues every day until the Spine-tail finds a mate, when all at once it becomes silent. The nest is placed in a low thorn-bush, sometimes only two or three feet above the ground, and is an oblong structure of sticks, twelve or fourteen inches in depth, with the entrance near the top, and reached by a tubular passage made of slender sticks, and six or seven inches long. From the top of the nest a crooked passage leads to the cavity near the bottom; this is lined with a little fine grass, and nine eggs are laid, pear-shaped and pale bluish white in colour. I have found several nests with nine eggs, and therefore set that down as the full number of the clutch, though I confess it seems very surprising that this bird should lay so many. When the nest is approached, the parent birds remain silent and concealed at some distance. When the nest is touched or shaken, the young birds, if nearly fledged, have the singular habit of running out and jumping to the ground to conceal themselves in the grass.

I have no doubt that this species varies greatly in its habits in different districts, and probably also in the number of eggs it lays. Mr. Barrows, an excellent observer, says it lays three or four light blue eggs. He met with it at Concepcion, in the northern part of the Argentine Republic, and writes that it is "an abundant species in thorny hedges or among the masses of dwarfed and spiny bushes, which cling to each other so tenaciously amid the general desolation of the sandy barrens." The nests which he describes vary also in some particulars from those I have seen. "Entrance is gained by the bird," he says, "through a long tube, which is built on to the nest at a point about half way up the side. This tube is formed by the interlocking of thorny twigs, and is supported by the branches and twigs about it. It may be straight or curved; its diameter externally varies from two to four inches, and its length from one to two feet. The passage-way itself is but just large enough to admit the birds one at a time, and it has always been a mystery to me how a bird the size of a Chipping-Sparrow could find its way through one of these slender tubes, bristling with thorns, and along which I found it difficult to pass a smooth slender twig for more than five or six inches. Yet they not only pass in and out easily, but so easily that I was never yet able to surprise one in the nest, or to see the slightest disturbance of it by the bird's hurried exit."

The bird has a very wide range in South America, and Mr. Salmon observed its breeding-habits in Antioquia in Colombia. There also the bird varies the form of its nest, making it as large as that of an English Magpie, and roofing the top with a mass of large leaves to protect it from the heavy rains. The eggs, he says, are very pale greenish blue, nearly white; but he does not give the number.

---

195. SYNALLAXIS WHITII, Scl.

(WHITE'S SPINE-TAIL.)

Synallaxis whitii, *Scl. Ibis,* 1881, p. 600, pl. xvii. fig. 2; *White, P. Z. S.* 1882, p. 611 (Salta).

Description.—Above dark brown, lores and superciliaries white; wings and tail chestnut-red; beneath cinnamomeous, paler on the middle of the belly; large spot in the middle of the throat black; inner margins of wing-feathers and under wing-coverts cinnamomeous; bill black, feet pale hazel: whole length 5·7 inches, wing 2·5, tail 2·6.

Hab. Northern Argentina.

White obtained a single example of this new species near Oran in November 1880. It is most nearly allied to *S. scutata* of Brazil.

---

196. SYNALLAXIS PHRYGANOPHILA (Vieill.).

(PRETTY-THROATED SPINE-TAIL.)

Synallaxis phryganophila, *Burm. La-Plata Reise,* ii. p. 469 (Paraná); *Scl. et Salv. Nomencl.* p. 64; *Scl. P. Z. S.* 1874, p. 17; *White, P. Z. S.* 1882,

p. 611 (Buenos Ayres); *Barrows, Bull. Nutt. Orn. Cl.* viii. p. 207 (Entrerios).

Description.—Above, front brown, crown chestnut, superciliaries white; sides of head, neck, back, and tail pale fulvous brown, with broad blackish striations on the neck and back; upper wing-coverts pale chestnut; wing-feathers blackish, the outer webs edged with pale fulvous brown; beneath, upper half of throat sulphur-yellow, lower half black, with a white patch on each side of the black; breast and belly whitish, washed with earthy brown, slightly fulvous on the breast and flanks; under wing-coverts fulvous white; bill and feet horn-colour: whole length 8·5 inches, wing 2·4, tail 4·6.

Hab. Paraguay, Argentina, and Uruguay.

This pretty Spine-tail is nowhere common in the Argentine country, and in Buenos Ayres it is exceedingly scarce. It is rather large for a *Synallaxis,* the total length being nine inches. The two middle feathers of the acuminated tail greatly exceed the others in length, measuring five inches. The plumage is very pale brown, marked with fuscous; the crown and wing-coverts rufous. The beauty of the bird is in the throat, which has three strongly contrasted colours, distinguishing it from all other *Synallaxes.* In the angle of the beak the colour is sulphur-yellow, under this is a patch of velvet-black, and on each side of the yellow and black a pure white patch.

Mr. Barrows has the following very interesting note on its nesting-habits:—"A nest containing four white eggs, faintly tinted with blue, was found in a thorny tree, and some eight feet from the ground. The nest was quite similar to the one just described (of *S. albescens*), but the cavity in which the eggs were laid was near the *top* of the body of the nest, while the passage-way descended from it to the base of the nest, and there becoming external rose gradually to the level of the eggs at a distance of almost three feet."

---

197. SYNALLAXIS STRIATICEPS (Lafr. et d'Orb.).

(STRIPED SPINE-TAIL.)

Synallaxis striaticeps, *Burm. La-Plata Reise,* ii. p. 469 (Paraná); *Hudson, P. Z. S.* 1872, p. 544 (Rio Negro); *Scl. et Salv. Nomencl.* p. 64; *Scl. P. Z. S.* 1874, p. 21; *Salvin, Ibis,* 1880, p. 358 (Salta); *White, P. Z. S.* 1883, p. 39 (Cordova); *Barrows, Bull. Nutt. Orn. Cl.* viii. p. 208 (Entrerios).

Description.—Above earthy brown, darker on the crown, which has slight greyish striations; broad superciliaries white; upper wing-coverts pale chestnut; wing-feathers blackish, glossed with olive; tail pale chestnut; beneath white; under wing-coverts pale fulvous; bill and feet horn-colour: whole length 5·9 inches, wing 2·4, tail 2·4. *Female* similar.

Hab. Bolivia, Argentina, Uruguay, and Patagonia.

This species has a wide range south of the Equator, being found in Bolivia, Uruguay, and throughout the Argentine Republic, including Patagonia. In its habits it differs widely from other *Synallaxes*, and in structure and coloration is also unlike its relatives.

The beak is longer and more curved, the claws more crooked, and the tail stiffer than in other *Synallaxes*, and this difference in structure corresponds to a different mode of life. The Striped Spine-tail creeps on the trunk and larger branches of trees, seeking its insect-food in the crevices of the bark, and when seen clinging to the trunk, supported by its tail in a vertical position, with head thrown far back, and progressing upwards by short quick hops, it looks wonderfully like a small *Picolaptes* with shortened beak. It is very restless, and while searching for insects constantly utters a short, trilling, querulous-sounding note.

It builds an open nest in the fork of a branch, of soft grasses and hair, thickly lined with feathers, and lays four or five pure white eggs.

---

198. SYNALLAXIS ORBIGNII, Reichenb.

(D'ORBIGNY'S SPINE-TAIL.)

Synallaxis orbignii, *Scl. et Salv. Nomencl.* p. 64; *Scl. P. Z. S.* 1874, p. 22, et 1879, p. 461; *Salvin, Ibis,* 1880, p. 358 (Salta). **Synallaxis humicola,** *Burm. La-Plata Reise,* ii. p. 468 (Mendoza, Paraná). **Synallaxis crassirostris,** *Leybold, J. f. O.* 1865, p. 401. **Synallaxis flavigularis,** *Döring, Exp. al Rio Negro, Zool.* p. 45.

Description.—Above pale earthy brown, superciliaries whitish; wing-feathers blackish, the basal part of the secondaries pale chestnut, forming a transverse band; four middle tail-feathers blackish, the lateral pale chestnut; beneath whitish brown; middle of throat pale rufous; under wing-coverts fulvous; bill and feet horn-colour: whole length 6·5 inches, wing 2·2, tail 3·2.

Hab. Bolivia and Argentina.

This Spine-tail has been noticed by various observers in Northern Argentina, and, if we are correct in referring *S. flavigularis* of Döring to the same species, occurs also in the Rio Negro district.

---

199. SYNALLAXIS MODESTA, Eyton.

(MODEST SPINE-TAIL.)

Synallaxis modesta, *Scl. et Salv. Nomencl.* p. 64; *Scl. P. Z. S.* 1872, p. 544 (Rio Negro), et 1874, p. 23. **Synallaxis flavogularis,** *Burm. La-Plata Reise,* ii. p. 468 (Mendoza).

Description.—Above earthy brown, superciliaries whitish; wings blackish, the basal half of secondaries pale clear brown, forming a transverse bar; tail black, the outer webs of all the rectrices pale chestnut; beneath whitish brown, with a large fulvous spot on the throat; under wing-coverts fulvous; bill and feet pale brown: whole length 6·0 inches, wing 2·4, tail 2·5. *Female* similar.

Hab. Argentina, Patagonia, Chili.

This species so closely resembles the following in size and dull earthy-brown colour, that where seen in the thickets it is impossible to distinguish them. In habits they also seem alike; but this bird is, I think, less retiring, for I have seen it associating with other species of *Synallaxis*.

On comparing specimens together, however, it is easy to separate the present bird from *S. sordida* by noticing the colour of the external rectrices, which are black, externally edged with rufous, instead of being wholly rufous.

---

200. SYNALLAXIS SORDIDA, Less.
(SORDID SPINE-TAIL.)

Synallaxis sordida, *Hudson, P. Z. S.* 1872, p. 543 (Rio Negro); *Scl. et Salv. Nomencl.* p. 64; *Scl. P. Z. S.* 1874, p. 23; *Durnford, Ibis,* 1878, p. 396 (Centr. Patagonia); *White, P. Z. S.* 1883, p. 39 (Cordova); *Barrows, Bull. Nutt. Orn. Cl.* viii. p. 208 (Entrerios).

Description.—Above earthy brown; wing-feathers blackish brown, their basal parts pale chestnut-brown, forming a transverse bar; tail blackish, the three outer rectrices and outer web of the fourth rectrix on each side wholly pale chestnut-brown; beneath pale earthy brown, clearer on the belly, and with a bright fulvous spot on the throat; under wing-coverts pale cinnamomeous; bill and feet blackish: whole length 6·5 inches, wing 2·3, tail 2·8. *Female* similar.

Hab. Patagonia and East Argentina.

This species, which, on close comparison, is at once distinguishable from *S. modesta* by the absence of any black colour on the three exterior pairs of tail-feathers, ranges from the extreme north of the Argentine Republic to Patagonia, where it is quite common, and is invariably found in dry situations abounding in thorny vegetation.

It does not migrate, and lives with its mate in thorny bushes, but does not attempt to conceal itself, and sits much on the summit of a bush, where the male in spring utters at intervals a clear trilling call. In its inactive disposition, slow deliberate movements, also in its language, it strikingly resembles the *Phacellodomus ruber*. In its nidification it also comes nearest to that species. The nest is a large structure of sticks, eighteen inches to two feet long, placed upright among the twigs at the summit of a bush. From the top where the entrance is placed, a winding passage leads down to the chamber at the bottom of the nest; this is lined with soft dry grass and feathers, and four pure white eggs are laid.

---

201. SYNALLAXIS SULPHURIFERA, Burm.

(YELLOW-MARKED SPINE-TAIL.)

Synallaxis sulphurifera, *Hudson, P. Z. S.* 1872, p. 544 (Rio Negro); *Scl. et Salv. Nomencl.* p. 64; *Scl. P. Z. S.* 1874, p. 24; *Durnford, Ibis,* 1877, p. 180, et 1878, p. 61 (Buenos Ayres); *Barrows, Bull. Nutt. Orn. Cl.* viii. p. 208 (Entrerios).

Description.—Above brown, slightly olivaceous; wings blackish; lesser wing-coverts, margins of the feathers of the greater wing-coverts, and outer webs of the basal halves of the wing-feathers pale chestnut; tail-feathers chestnut-brown, ends much elongated and pointed; beneath white, throat and breast mottled with grey, spot in middle of throat sulphur-yellow; flanks washed with brown; bend of wing and under wing-coverts fulvous; bill and feet pale horn-colour: whole length 6·3 inches, wing 2·1, tail 3·0. *Female* similar.

Hab. South Argentina and Patagonia.

I have found this *Synallaxis,* which was first described by Prof. Burmeister, from specimens obtained near Buenos Ayres, in the swamps along the Plata river; also on the Rio Negro, in Patagonia, where, however, it is very scarce. It inhabits the dense rush-beds growing in the water, where the *Limnornis curvirostris* is also found. It closely resembles that species in habits and language, and also assimilates to it in colour and in the rather long, curved beak, sharp claws, stout body, and short stiff tail. It is stationary, pairs for life, and lives always closely concealed in its chosen bed of close-growing rushes. When a person approaches their hiding-place the two birds creep up to the summit of the rushes, protesting in peculiar, loud, angry rattling notes. The *Limnornis,* which also pairs for life, has precisely the same habit.

Durnford describes the nest, found in a rush-bed, as a circular or domed structure of grass, with the aperture at the side; the eggs white.

---

202. SYNALLAXIS PATAGONICA (Lafr. et d'Orb.).

(PATAGONIAN SPINE-TAIL.)

Synallaxis patagonica, *Hudson, P. Z. S.* 1872, p. 544 (Rio Negro); *Scl. et Salv. Nomencl.* p. 64; *Scl. P. Z. S.* 1874, p. 24; *Durnford, Ibis,* 1877, p. 35 (Chupat).

Description.—Above greyish earthy brown; wing-feathers blackish brown, basal halves of secondaries very pale clear brown, forming a transverse band; tail blackish, edged with greyish brown; outer web of outer feather on each side pale brown; beneath cinereous, with an obscure blackish spot on the throat; belly and flanks dull buff; under wing-coverts cinnamomeous; bill and feet blackish: whole length 6·0 inches, wing 2·2, tail 2·5. *Female* similar.

Hab. Patagonia.

This dull-coloured little bird, which is found in Patagonia and also near the Andes in the north-western provinces of the Argentine Republic, is one of those species which diverge greatly in habits from the typical *S. ruficapilla* and its nearest allies. The body is stout, the tail, square and short, is carried vertically as in the House-Wren.

The Patagonian Spine-tail is a resident in the Rio Negro district. It is a silent, shy, solitary little bird, which lives on the ground and seeks its food after the manner of the Cachalote (*Homorus*). Being small and feeble, however, it does not hunt about the roots of trees and large bushes like the larger and more powerful *Homorus*, but keeps under the diminutive scrubby plants in open sterile situations. About the roots of these wiry little bushes, only twelve to eighteen inches high, the bird searches for small insects, and when disturbed has a feeble jerky flight, which carries it to a distance of about twenty yards. It flies with great reluctance, and when approached runs swiftly away, leaving a person in doubt as to whether he has seen a mouse or a little obscure bird. The only note I have heard it utter is a faint creaking sound when alarmed or flying.

---

203. SYNALLAXIS HUDSONI, Scl.

(HUDSON'S SPINE-TAIL.)

Synallaxis hudsoni, *Scl. P. Z. S.* 1874, p. 25; *Durnford, Ibis,* 1877, p. 36, et 1878, p. 396 (Chupat, Central Patagonia); *Barrows, Bull. Nutt. Orn. Cl.* viii. p. 209 (Azul, Sierra de la Ventana). **Synallaxis sclateri,** *Cab. J. f. O.* 1878, p. 196.

Description.—Above fulvous brown, mottled with black, each feather being marked with a large black spot; on the upper part of the back the feathers are faintly edged with whitish grey; wings blackish, basal halves of feathers pale clear brown, forming a transverse bar, the terminal part of the feathers slightly edged on the outer webs and tips with ochraceous; tail blackish, the outer pair of rectrices and broad tips of the next two pairs on each side very pale brown, the two middle feathers broadly margined on both webs with pale greyish brown; beneath pale ochraceous brown, with a pale sulphur-yellowish gular spot; flanks with a few black marks; under wing-coverts light cinnamon; bill and feet pale horn-colour: whole length 7·8 inches, wing 3·2, tail 3·5. *Female* similar.

Hab. Argentina.

This Spine-tail, which Sclater has named after me, is the Argentine representative of *S. humicola* of Chili. It is common on the pampas, and is sometimes called by the gauchos *"Tiru-riru del campo,"* on account of its resemblance in the upper plumage and in language to *Anumbius acuticaudatus*, which is named *"Tiru-riru,"* in imitation of its call-note. The addition of *del campo* signifies that it is a bird of the open country. It is, in fact, found exclusively on the grassy pampas, never perching on trees, and in habits is something like a Pipit, usually being taken for one when first seen. It is quite common everywhere on the pampas, and specimens have also been obtained in Cordova, Uruguay, and Patagonia.

This Spine-tail is resident, solitary, and extremely timid and stealthy in its movements, living always on the ground among the long grass and cardoon-thistles. At times its inquisitiveness overcomes its timidity, and the bird then darts up three or four yards into the air, and jerking its tail remains some moments poised aloft with breast towards the intruder, emitting sharp little notes of alarm, after which it darts down again and disappears in the grass. When driven up it has a wild zigzag flight, and after reaching a considerable height in the air darts down again with astonishing swiftness, and comes back to the very spot from which it rose. It is, however, incapable of sustained flight, and after being flushed two or three times refuses to rise again. In spring the male perches on the summit of a cardoon-bush, or other slight elevation, and at regular intervals utters a pleasing and melancholy kind of song or call, which can be heard distinctly at a distance of a thousand yards, composed of four long clear plaintive notes, increasing in strength, and succeeded by a falling trill. When approached it becomes silent, and dropping to the ground conceals itself in the grass. Under a cardoon-bush or tussock of grass it scoops out a slight hollow in the ground, and builds over this a dome of fine dry grass, leaving a small aperture arched like the door of a baker's oven. The bed is lined with dry powdered horse-dung, and the eggs are five, bluntly pointed and of a very pale buff colour. The interior of the nest is so small that when the five young birds are fledged they appear to be packed together very closely, so that it is difficult to conceive how the parent bird passes in and out.

The nest is always very cunningly concealed, and I have often spent days searching in a patch of cardoon-bushes where the birds were breeding without being able to find it.

---

204. SYNALLAXIS MALUROIDES, d'Orb.

(WREN-LIKE SPINE-TAIL.)

Synallaxis maluroides, *Scl. et Salv. Nomencl.* p. 64; *Scl. P. Z. S.* 1874, p. 26; *Durnford, Ibis*, 1877, p. 180, et 1878, p. 61 (Buenos Ayres); *Barrows, Bull. Nutt. Orn. Cl.* viii. p. 208 (Entrerios).

Description.—Above, front and middle of crown chestnut; hind head, neck, and back pale fulvous brown, thickly marked with longitudinal black shaft-spots; lores white; wings blackish, the feathers edged with pale ochraceous, the basal part of secondaries very pale brown, forming a transverse bar; tail pale chestnut-brown, the two middle feathers with a broad black mark on the inner web; beneath white, breast and flanks washed with pale brown, and freckled with very small dark brown spots; under wing-coverts white; bill and feet pale horn-colour: whole length 6·1 inches, wing 2·0, tail 2·9. *Female* similar.

Hab. South Argentina.

D'Orbigny discovered this small Spine-tail near Buenos Ayres city, but did not record its habits. Like the species just described it is abundant on the pampas, but in its habits resembles a Wren of the genus *Cistothorus* rather than a Pipit, being partial to moist situations, where there is a rank growth of grass and herbage. The wings are very short, and the flight so feeble that the bird refuses to rise after being pursued a distance of one or two hundred yards. And yet I am not prepared to say that it does not migrate, as I have found that in spring it all at once becomes very abundant, while in the cold season it is rarely seen. It is solitary, and in spring sits on a thistle or stalk, uttering at short intervals its small grasshopper-like song or call. The nest is a slight open structure of grass, lined with a few feathers, placed in a tuft of grass or reeds. The eggs are pure white in colour.

---

205. CORYPHISTERA ALAUDINA, Burm.

(LARK-LIKE CORYPHISTERA.)

Coryphistera alaudina, *Burm. La-Plata Reise*, ii. p. 470 (Paraná); *Scl. et Salv. Nomencl.* p. 64; *Salv. Ibis*, 1880, p. 359 (Tucuman, Salta); *White, P. Z. S.* 1883, p. 40 (Cordova).

Description.—Above dark greyish brown; elongated crest-feathers blackish; ear-coverts chestnut; on the back, upper tail-coverts, and upper wing-coverts the feathers have white and whity-brown edgings; wing-feathers blackish, the basal part of the inner webs pale brown; rectrices bright chestnut, broadly tipped with blackish; beneath white, thickly striated with fulvous brown; under wing-coverts pale cinnamon; bill and feet light brown: whole length 6·2 inches, wing 2·7, tail 2·3. *Female* similar.

Hab. Argentina.

This highly interesting little bird, the only known member of its genus, inhabits the dry plains of Paraná and Cordova.

The following meagre note from White, which only serves to excite curiosity, comprises all that we know of its habits:—

"These birds are not found in dense woods, but in the open, tenanted only by a few small trees or bushes. Five or six are usually seen running about together with a quick, abrupt movement, meanwhile uttering a sharp cry."

---

206. ANUMBIUS ACUTICAUDATUS (Less.).

(FIREWOOD-GATHERER.)

Anumbius acuticaudatus, *Burm. La-Plata Reise,* ii. p. 467 (Paraná, Mendoza); *Scl. et Salv. Nomencl.* p. 64; *Hudson, P. Z. S.* 1874, p. 159 (Buenos Ayres); *Durnford, Ibis,* 1877, p. 181 (Buenos Ayres); *White, P. Z. S.* 1882, p. 612 (Buenos Ayres); *Barrows, Bull. Nutt. Orn. Cl.* viii. p. 211 (Entrerios).

Description.—Above earthy brown, forehead chestnut, superciliaries white, head, neck, and back marked with black striations; primaries blackish, secondaries pale chestnut-brown; tail black, all the feathers except the middle pair broadly tipped with cream-colour; beneath pale ochraceous brown, white on the throat, the white bordered on each side by numerous small black spots; bill and feet pale horn-colour: whole length 8·3 inches, wing 3·6, tail 3·7. *Female* similar.

Hab. Argentina, Uruguay, Paraguay.

This is a common and very well-known species throughout the Argentine country and Patagonia, also in Uruguay and Paraguay, and is variously called *Espinero* (Thorn-bird), *Tiru-riru,* in imitation of its note, and *Añumbi* (the Guarani name); but its best known name is *Leñatero,* or "Firewood-Gatherer," from the quantity of sticks which it collects for building-purposes.

The Firewood-Gatherer is a resident in Argentina, and pairs for life. Sometimes the young birds remain with their parents for a period of three or four months, all the family going about and feeding in company, and roosting together in the old nest. The nest and the tree where it is placed are a favourite resort all the year round. Here the birds sit perched a great deal, and repeat at intervals a song or call, composed of four or five loud ticking chirps, followed by a long trilling note. They feed exclusively on the ground, where they creep about, carrying the body horizontally and intently searching for insects. When disturbed, they hurry to their usual refuge, rapidly beating their very feeble wings, and expanding the broad acuminated tail like a fan. When the male and female meet at their nest, after a brief separation, they sing their notes in concert, as if rejoicing over their safe reunion; but they seldom separate, and Azara says that when one incubates, the other sits at the entrance to the nest, and that when one returns to the nest with food for the young the other accompanies it, though it has found nothing to carry.

To build, the *Añumbi* makes choice of an isolated tree in an open situation, and prefers a dwarf tree with very scanty foliage; for small projecting twigs and leaves hinder the worker when carrying up sticks. This is a most laborious operation, as the sticks are large and the bird's flight is feeble. If the tree is to its liking, it matters not how much exposed to the winds it may be, or how close to a human habitation, for the bird is utterly unconcerned by the presence of man. I have frequently seen a nest in a shade or ornamental tree within ten yards of the main entrance to a house; and I have also seen several on the tall upright stakes of a horse-corral, and the birds working quietly, with a herd of half-wild horses rushing round the enclosure beneath them, pursued by the men with lassos. The bird uses large sticks for building, and drops a great many; frequently as much fallen material as would fill a barrow lies under the tree. The fallen stick is not picked up again, as the bird could not rise vertically with its load, and is not intelligent enough, I suppose, to recover the fallen stick, and to carry it away thirty yards from the tree and then rise obliquely. It consequently goes far afield in quest of a fresh one, and having got one to its liking, carefully takes it up exactly by the middle, and, carrying it like a balancing-pole, returns to the nest, where, if one end happens to hit against a projecting twig, it drops like the first. The bird is not discouraged, but, after a brief interview with its mate, flies cheerfully away to gather more wood.

Durnford writes wonderingly of the partiality for building in poplar trees shown by this bird in Buenos Ayres, and says that in a tall tree the nest is sometimes placed sixty or seventy feet above the ground, and that the bird almost invariably rises with a stick at such a distance from the tree as to be able just to make the nest, but that sometimes failing it alights further down, and then climbs up the twigs with its stick. He attributes the choice of the tall poplar to *ambition*; but the *Añumbi* has really a much simpler and lowlier motive. In the rich Buenos Ayrean soil all trees have a superabundance of foliage, and in the slim poplar alone can the nest be placed where the bird can reach it laden with building-material, without coming in contact with long projecting twigs.

The nest of the *Añumbi* is about two feet in depth, and from ten to twelve inches in diameter, and rests in an oblique position amongst the branches. The entrance is at the top, and a crooked or spiral passage-way leads down to the lower extremity, where the breeding-chamber is situated; this is lined with wool and soft grass, and five white eggs are laid, varying considerably in form, some being much more sharply pointed than others.

The nest, being so secure and comfortable an abode, is greatly coveted by several other species of birds to breed in; but on this subject I have already spoken in the account of the genus *Molothrus*. When deprived of their nest, the birds immediately set to work to make a new one; but, often enough, without being ejected from the first they build a second nest, sometimes demolishing the first work to use the materials. I watched one pair make three nests before laying; another pair made two nests, and after the second was completed they returned to the first and there elected to remain. Two or three nests are sometimes seen on one tree, and Azara says he has seen as many as six. Mr. Barrows observed the bird at Concepcion, where it is very common, and writes that in that district the nest is sometimes four feet long with an average diameter of two feet, and that the same nest in some cases is used for several seasons successively; also that several nests are sometimes joined together and all occupied at the same time.

---

207. LIMNORNIS CURVIROSTRIS, Gould.

(CURVED-BILL RUSH-BIRD.)

Limnornis curvirostris, *Scl. et Salv. Nomencl.* p. 64; *Durnford, Ibis,* 1877, p. 182 (Buenos Ayres).

Description.—Above rufous-brown, brighter on the rump; lores and superciliaries white; wings and tail chestnut-brown; beneath white; flanks and under tail-coverts pale brown; under wing-coverts white; bill and feet horn-colour: whole length 7·0 inches, wing 3·0, tail 2·0. *Female* similar.

Hab. Argentina and Uruguay.

This species is found everywhere in marshy places in the eastern part of the Argentine Republic, and is also common in Uruguay, where

Darwin discovered it. It inhabits dense reed-beds which grow in the water, and is not found in any other situation. It pairs for life, has a very feeble flight, and flies with great reluctance, but lives always in close concealment in one spot. It is, however, very inquisitive, and when approached the two birds creep up to the summit of the rushes and utter peculiar loud, rattling, and jarring notes, as if angrily protesting against the intrusion.

The Rush-bird has a stout body and short graduated tail, strong claws, and a slender curved beak three-fourths of an inch long. The upper plumage is brown, the tail rufous, the under surface and a mark over the eye white.

---

208. PHACELLODOMUS FRONTALIS (Licht.).

(RED-FRONTED THORN-BIRD.)

Phacellodomus frontalis, *Burm. La-Plata Reise,* ii. p. 467 (Tucuman); *Scl. et Salv. Nomencl.* p. 65; *Salvin, Ibis,* 1880, p. 359 (Salta); *White, P. Z. S.* 1882, p. 612 (Buenos Ayres). **Phacellodomus sincipitalis,** *White, P. Z. S.* 1883, p. 433.

Description.—Above nearly uniform olive-brown, crown blackish brown, superciliaries white; beneath dirty brownish white; under wing-coverts pale cinnamon; bill and feet horn-colour: whole length 6·8 inches, wing 2·6, tail 2·6. *Female* similar.

Hab. S. America, from Venezuela to Argentina.

The Red-fronted Thorn-bird, which is found in the Northern provinces of Argentina, and only occurs as a straggler near Buenos Ayres, resorts to the thickets, and in its habits is said to resemble the *Synallaxes* of the group to which *S. spixi* and *S. albescens* belong. It builds a large nest of sticks, and White says that it makes a peculiar chattering sound that has the effect of exciting other small birds, and causes them to crowd about it.

---

209. PHACELLODOMUS SIBILATRIX, Scl.

(WHISTLING THORN-BIRD.)

Phacellodomus sibilatrix, *Scl. P. Z. S.* 1879, p. 461; *White, P. Z. S.* 1882, p. 612 (Buenos Ayres); *Barrows, Bull. Nutt. Orn. Cl.* viii. p. 209 (Entrerios).

Description.—Above dull olive-brown, paler on the sides of the head; front and lesser wing-coverts chestnut-red; wing-feathers blackish, edged with olive-brown; tail light chestnut, the middle pair of feathers and the inner webs of the two next pairs brown, like the back; beneath dirty brownish white, washed with ochraceous on the flanks; under wing-coverts bright cinnamon: whole length 5·3 inches, wing 2·2, tail 2·2.

Hab. Argentina.

This species resembles *P. frontalis*, but differs in its much smaller size, and in having the upper lesser wing-coverts bright rufous.

It inhabits the thorny woods of the northern districts of the Argentine country, but I have no reason to regret that I have not personally observed this species, since Mr. Barrows's careful account of its nesting-habits leaves nothing to be desired. He writes:—"An abundant species among the open woods along the Uruguay, and hardly distinguishable at ten paces from half a dozen others. Its nest, however, is unmistakable. The birds begin by fixing a few crooked and thorny twigs among the terminal sprays of some slender branch which juts out horizontally from a tree, or rises obliquely from near its base, and around these twigs as a nucleus more are gathered, until, by the time the nest has reached the proper size, its weight has bent the branch so that its tip points directly to the earth. Nests which are thus begun at a distance of fifteen or twenty feet from the ground are often only two or three feet from it when finished, and a thorough soaking by a heavy rain will sometimes weigh them down until they actually touch. They are more or less oval or cylindrical in shape, and commonly about two feet long by twelve or fifteen inches in diameter, and contain from a peck to a bushel of twigs and thorns. The nest-cavity within is small in proportion to the size of the nest, and occupies its upper part. It is reached by a more or less direct passage-way from below, the external opening being very nearly at the lowest part of the nest, though sometimes a little shelf, or even a pocket, is built on to the side, forming a resting-place in front of the door.

"The nests vary interminably in size and shape, but are pretty constant in the material used; this being almost always irregular and thorny twigs of various trees growing in the neighbourhood, while the interior is formed of less thorny twigs and with some wool and hair. Usually, also, if the material be at hand, a quantity of old dry horse-droppings is placed loosely on the top of the nest, and gradually becomes felted into it, rendering it more nearly waterproof. In place of this I have frequently found quantities of broken straw, weed-stalks, grass, and even chips; all doubtless collected from the ridges of drift which the last overflow of the river had left near at hand. So compactly is the whole nest built, that it often lasts more than one year, and may sometimes serve the same pair two successive summers. More often, however, a new nest is built directly above the old one, which serves as a foundation, and occasionally as many as three nests are seen thus on the same branch-tip, two of them at least being occupied. When other branches of the same tree are similarly loaded, and other trees close at hand also bear the same kind of fruit, the result is very picturesque. The eggs, which are white, are laid from October 1 to January 1, but many of the birds work at nest-building all the winter, sometimes spending months on a single nest."

210. PHACELLODOMUS STRIATICOLLIS (d'Orb. et Lafr.).

(RED-WINGED THORN-BIRD.)

Anumbius striaticollis, *d'Orb. Voy., Ois.* p. 255 (Buenos Ayres). **Phacellodomus striaticollis,** *Scl. et Salv. Nomencl.* p. 65.

Description.—Above, forehead and crown rufous ; lores white ; sides of head, neck, and back brown ; whole wing chestnut, the feathers tipped with blackish ; tail chestnut ; beneath white, clearer on the throat and breast, which are marked with slight white shaft-spots ; sides of breast and flanks washed with reddish brown ; under tail-coverts brown ; under wing-coverts cinnamon ; bill and feet horn-colour : whole length 8·0 inches, wing 2·9, tail 3·1. *Female* similar.

Hab. Argentina and Uruguay.

D'Orbigny, who discovered this species in Uruguay, and found it also near Buenos Ayres, remarks that it lives in pairs in the thorny bushes, and resembles its congeners in habits. It is, however, certainly not common in the vicinity of the Argentine capital, for Hudson has never met with it.

211. PHACELLODOMUS RUBER (Vieill.).

(RED THORN-BIRD.)

Phacellodomus ruber, *Burm. La-Plata Reise,* ii. p. 467 (Paraná, Cordova) ; *Scl. et Salv. Nomencl.* p. 65 ; *Durnford, Ibis,* 1877, p. 183 (Buenos Ayres) ; *Barrows, Bull. Nutt. Orn. Cl.* viii. p. 210 (Entrerios). **Phacellodomus maculipectus,** *Cab. J. f. O.* 1883, p. 109 (?).

Description.—Above olive-brown, front chestnut ; tail brownish chestnut ; beneath whitish, throat, breast, and flanks washed and mottled with bright reddish brown ; under wing-coverts and inner margins of wing-feathers bright cinnamon ; bill and feet horn-colour : whole length 7·3 inches, wing 2·6, tail 3·2. *Female* similar.

Hab. Argentina.

This is a common species throughout the eastern portion of the

$\frac{5}{6}$

HOMORUS LOPHOTES.

Argentine country, and extends as far south as the southern boundary of the Buenos Ayrean province.

It is resident, living in pairs in places where there are scattered thorny trees and bushes, and is never found in deep woods. It never attempts to conceal itself, but, on the contrary, sits exposed on a bush and will allow a person to approach within three or four yards of it. Nor has it the restless manner of most Synallaxine birds which live in the same places with it, but moves in a slow deliberate way, and spends a great deal of time sitting motionless on its perch, occasionally uttering its call or song, composed of a series of long, shrill, powerful notes in descending scale and uttered in a very leisurely manner. It builds a large oblong nest of sticks, about two feet deep, and placed obliquely among the thorny twigs of a bush or low tree. Mr. Barrows writes:—"There are commonly two cavities in the nest, one being half open to the weather, and forming the entrance, the other further back and connected with the former by only a short passage-way, which in many cases is reduced to a simple hole through a broad partition, which alone separates them." The eggs are four and of a pure white.

The name commonly used for this species is founded on the "*Anumbé roxo*" of Azara's 'Apuntamientos'; but the description given there of the bird's nesting-habits shows either that some other species was meant —perhaps *P. sibilatrix*, Döring—or that the nesting-habits of a different bird have been assigned to *P. ruber*.

---

212. HOMORUS LOPHOTES, Reichenb.

(BROWN CACHALOTE.)

[Plate IX.]

Homorus lophotes, *Reichb. Handb.* p. 172; *Barrows, Bull. Nutt. Orn. Cl.* vol. viii. p. 212 (Entrerios); *Hudson, Ibis,* 1885, p. 283 (Buenos Ayres). **Anabates unirufus,** *Burm. La-Plata Reise,* ii. p. 466 (Cordova). **Homorus unirufus,** *Scl. et Salv. Nomencl.* p. 65; *White, P. Z. S.* 1882, p. 612 (Catamarca).

Description.—Above brown, tinged with olive on the back, but clear and rufescent on the hind head and rump; crest-feathers blackish brown; wings blackish; tail chestnut; beneath earthy brown, throat rufous; under wing- and tail-coverts and inner margins of wing-feathers pale rufous; bill pale bluish, feet bluish horn-colour: whole length 9·8 inches, wing 4·6, tail 4·2. *Female* similar.

Hab. Argentina.

This species interested me greatly during my observations of the

Argentine birds, but, owing to its rarity and to its recluse habits, my knowledge of it is very scanty. It is found in the north and north-western parts of the Argentine territory; in the province of Buenos Ayres its presence is confined to the narrow strip of subtropical wood fringing the low shores of the Plata river.

When surprised, its white eye, blue dagger-like beak, and raised crest give it a strikingly bold and angry appearance, the effect of which is heightened by the harsh rasping scream it utters when disturbed. This resentful look is deceptive, however, for the bird is the shiest creature imaginable. Its language has the shrill excited character common to this most loquacious family; and at intervals throughout the day two birds, male and female, meet together and make the woods echo with their screaming concert. For many weeks after I had become familiar with these loud-sounding notes, while collecting in the littoral forest where it is found, the bird was still to me only a "wandering voice"; but I did not give up the pursuit till I had seen it several times and had also secured two or three specimens. I found one nest, though without eggs, a rough-looking domed structure, made with material enough to fill a barrow. I also discovered that the bird feeds exclusively on the ground, close to the boles of low-branching trees, where there is usually an accumulation of fallen bark, dead leaves, and other rubbish. Here the bird digs with its sharp beak for the small insects it preys on. When approached it does not fly away, but runs swiftly to the nearest tree, behind the trunk of which it hides, then scuttles on to the next tree, and so escapes without showing itself.

Mr. Barrows, who observed the Cachalote at Concepcion, says that it is a bird which cannot be overlooked, with an outrageous disposition and voice, and a nest the size of a barrel. He gives the following account of its nidification:—"His nest is built entirely of sticks, and many of them of goodly size, frequently as large round as your little finger and two feet or more long. These are disposed in such a way as to form a structure three or four feet in length by about two in breadth in the widest part, the whole very much resembling a gigantic powder-flask lying on its side among the lower branches of a spreading tree. It is quite loosely built and the nest-cavity is rather indefinite, being any portion of the floor of the nest which the bird selects for the reception of the eggs. These are usually three or four in number, pure white, and are laid from October until January. They can usually be counted through the loose floor of the nest, though sometimes its thickness prevents this."

213. HOMORUS GUTTURALIS (d'Orb. et Lafr.).

(WHITE-THROATED CACHALOTE.)

Anabates gutturalis, *Burm. La-Plata Reise*, ii. p. 467 (Mendoza). **Homorus gutturalis,** *Scl. et Salv. Nomencl.* p. 65; *Hudson, P. Z. S.* 1872, p. 545 (Rio Negro); *id. Ibis*, 1885, p. 284; *Durnford, Ibis*, 1877, p. 36, et 1878, p. 396 (Chupat, Central Patagonia).

Description.—Nearly uniform earthy grey, faintly tinged with olivaceous brown above, and much paler beneath; lores and upper part of throat pure white, lower part of throat black, or white and black mixed; under wing-coverts white, faintly tinged with pale cinnamon; beak and feet bluish grey: whole length 9·4 inches, wing 4·0, tail 3·2. *Female* similar.

Hab. Patagonia.

I found this bird quite common on the dry open plains in the neighbourhood of the Rio Negro in Patagonia. In size, form, and crest it is like the northern Cachalote, but has a white throat, while the rest of the plumage is of a pale earthy brown instead of rufous. Like the Brown Cachalote it is also shy in disposition, and, being so dull in colour and without the bright beak and eye tints, has not the bold, striking appearance of that species; still I do not think any ornithologist can meet with it and fail to be strongly impressed with its personality, if such a word can be applied to a bird.

Dendrocolaptine birds are, as a rule, builders of big nests and very noisy; *H. gutturalis* is, I believe, the loudest screamer and greatest builder of the family. Male and female live together in the same locality all the year; the young, when able to fly, remain with their parents till the breeding-season, so that the birds are found occasionally in pairs, but more frequently in families of five or six individuals. When feeding they scatter about, each bird attaching itself to a large bush, scraping and prodding for insects about the roots; and at intervals one of the old birds, ascending a bush, summons the others with loud shrill cries, on which they all hurry to the place of meeting, and from the summits of the bushes burst forth in a piercing chorus, which sounds at a distance like screams of hysterical laughter. At one place, where I spent some months, there were some bushes over a mile and a quarter from the house I lived in, where these birds used to hold frequent meetings, and in that still atmosphere I could distinctly hear their extravagant cries at this distance. After each performance they pursue each other, passing from bush to bush with a wild jerky flight, and uttering harsh angry notes.

They select a low, strong, wide-spreading bush to build in; the nest, which is made of stout sticks, is perfectly spherical and four to five feet deep, the chamber inside being very large. The opening is at the side near the top, and is approached by a narrow arched gallery, neatly made of slender sticks resting along a horizontal branch, and about fourteen inches long. This peculiar entrance, no doubt, prevents the intrusion of snakes and small mammals. So strongly made is the nest that I have stood on the dome of one and stamped on it with my foot without injuring it in the least, and to demolish one I had to force my gun-barrel into it, then prize it up by portions. I examined about a dozen of these enormous structures, but they were all met with before or after the laying-season, so that I did not see the eggs.

---

214. ANABAZENOPS OLEAGINEUS, Scl.

(OILY-GREEN ANABAZENOPS.)

Anabazenops rufo-superciliatus, *White, P. Z. S.* 1882, p. 613 (Catamarca). **Anabazenops oleagineus**, *Scl. P. Z. S.* 1883, p. 654.

Description.—Above and beneath dark olive-green; lores, superciliaries, and spots on the sides of the head yellowish white; wings blackish, the outer webs of the feathers olive-brown; tail bright chestnut; throat whitish yellow, on the lower part the yellow feathers tipped with olive; breast and belly spotted with same colour as the throat; bend of wing, under wing-coverts, and margins of inner webs of secondaries fulvous yellow; bill and feet blackish: whole length 6·2 inches, wing 3·2, tail 3·0. *Female* similar.

Hab. South Brazil and Argentina.

White obtained examples of this species on the Sierra of Totoral. He says it is a very wild bird and exceedingly scarce.

---

215. SITTOSOMUS ERITHACUS (Licht.).

(ROBIN-LIKE WOOD-HEWER.)

Sittosomus olivaceus, *White, P. Z. S.* 1882, p. 613 (Salta). **Sittasomus erithacus**, *Scl. et Salv. Nomencl.* p. 66.

Description.—Above olive-green, tinged with chestnut on the back, rump, and upper wing-coverts; wings black, the basal part of the inner webs of the secondaries fulvous yellow, forming a well-marked transverse bar; outer webs and broad tips of inner secondaries and whole of outer secondaries chestnut; tail and upper tail-coverts chestnut; beneath yellowish olive, brighter on the throat and breast; under wing-coverts fulvous yellow; under tail-coverts pale chestnut; bill and feet black: whole length 6·2 inches, wing 3·0, tail 3·0. *Female* similar.

Hab. South America from Colombia to Northern Argentina.

This is a straggler from the north, a specimen of which was obtained by White near Oran in 1880.

DRYMORNIS BRIDGESI.

216. GLYPHORHYNCHUS CUNEATUS (Licht.).

(WEDGE-BILLED WOOD-HEWER.)

Glyphorhynchus cuneatus, *Scl. et Salv. Nomencl.* p. 67; *White, P. Z. S.* 1882, p. 613 (Misiones).

Description.—Above olive-brown, superciliaries and small spots on the side of the head yellowish white, rump and upper tail-coverts chestnut; wings blackish, outer webs of wing-feathers olive-brown, basal part of inner webs of secondaries yellowish white, forming a transverse bar; tail chestnut; beneath earthy olive-brown, whitish yellow on the throat, and with spots of the same colour on the upper part of the breast; under wing-coverts white; bill and feet horn-colour: whole length 5·1 inches, wing 2·5, tail 2·4. *Female* similar.

Hab. South America from Colombia to Northern Argentina.

This is another northern form of which White obtained specimens in Misiones. He says it is not uncommon there in the thick woods, also in the orange-groves about the Jesuit ruins of St. Javier.

---

217. DENDROCOLAPTES PICUMNUS (Licht.).

(FLAT-BILLED WOOD-HEWER.)

Dendrocolaptes picumnus, *Scl. et Salv. Nomencl.* p. 67; *White, P. Z. S.* 1882, p. 613 (Misiones).

Description.—Above olivaceous brown; head blackish, thickly covered with yellowish-buff elongated shaft-spots; rump and upper tail-coverts tinged with chestnut; wing-feathers chestnut, tinged with olivaceous; tail chestnut; beneath pale earthy olive-brown, paler on the throat, the shafts of the feathers of the breast buffy white, forming long lines, the feathers of the belly and under tail-coverts transversely barred with blackish; under wing-coverts yellowish white, spotted with blackish; bill and feet black: whole length 10·5 inches, wing 4·7, tail 4·6. *Female* similar.

Hab. Brazil and Northern Argentina.

White obtained specimens of this species at Concepcion, "in the thickest parts of the woods, near the river, climbing up the trees, around which it turned in corkscrew fashion."

---

218. DRYMORNIS BRIDGESI, Eyton.

(BRIDGES'S WOOD-HEWER.)

[Plate X.]

Drymornis bridgesi, *Scl. et Salv. Nomencl.* p. 67; *White, P. Z. S.* 1882, p. 613 (Catamarca); *Barrows, Auk,* 1884, p. 20 (Entrerios). **Nasica gracilirostris,** *Burm. La-Plata Reise,* ii. p. 466 (Rio Quinto).

Description.—Above and below brown, brightest on the rump, and palest beneath; long superciliaries extending down the neck, and a mystacal stripe formed of white spots with faint black edgings; wing-feathers blackish; tail chestnut; on the throat a broad white band extending to the breast; breast and belly marked with large oblong white spots, which are margined with narrow black edgings; under wing-coverts and inner margins of secondaries bright cinnamon; bill and feet horn-colour: whole length 12·10 inches, wing 5·6, tail 4·6. *Female* similar.

Hab. North and West Argentina.

Eyton, when he described this species from Bridges's specimens, gave its habitat as Bolivia. It may inhabit the southern part of that Republic, but it is more probable that Bridges's examples were obtained in Northern Argentina, which he likewise visited. Bridges's Wood-hewer is the only member of its genus, and is one of the largest of the whole family *Dendrocolaptidæ*, measuring some 13 or 14 inches in length, inclusive of the great curved beak. Although found throughout the northern portion of the Argentine Republic, its habits are as yet imperfectly known, but the following extracts show that they must be very interesting, and that the bird is remarkably versatile. Mr. Barrows writes:—"These birds are somewhat gregarious, being oftenest seen in small parties of six to ten. They sometimes cling against the bark of a tree in the manner of Woodpeckers, but also spend much of their time on the ground. I think they use their curved bill much oftener for probing in the ground than for searching the bark of trees, as many of those shot had the base of the bill and the frontal feathers plastered with mud. In the stomach of the first one killed I found the silken sac, three fourths of an inch in diameter, or the eggs of a large spider, which makes holes ten or twelve inches deep in the hard soil everywhere."

White obtained examples of this species at Catamarca, and also notices its strangely contrasted habits. He writes:—"The cry of this bird is much the same as that of a Woodpecker, and it clings to the algarroba trees in a similar way; but in the afternoon it is seen scattered about on the sandy ground in the pursuit of insects."

---

219. XIPHOCOLAPTES ALBICOLLIS (Vieill.).

(WHITE-THROATED WOOD-HEWER.)

Xiphocolaptes albicollis, *Scl. et Salv. Nomencl.* p. 68; *White, P. Z. S.* 1882, p. 613 (Misiones).

Description.—Above, head black, with clear whitish-yellow shaft-spots; lores and long superciliaries white; neck, back, and upper wing-coverts olive-brown;

rump and upper tail-coverts washed with bright chestnut; wing-feathers dark chestnut, the outer webs glossed with olivaceous; tail chestnut; beneath pale olive-brown, buffy white on the throat and with similar shaft-spots on the breast; feathers of the belly and under tail-coverts transversely barred with black; under wing-coverts cinnamomeous yellow barred with black; bill and feet black: whole length 12·0 inches, wing 5·4, tail 4·8. *Female* similar.

Hab. Brazil and N. Argentina.

White obtained examples of this large Wood-hewer in Misiones.

---

220. XIPHOCOLAPTES MAJOR (Vieill.).

(CHESTNUT WOOD-HEWER.)

Xiphocolaptes major, *Scl. et Salv. Nomencl.* p. 68; *White, P. Z. S.* 1882, p. 613 (Salta); *Durnford, Ibis,* 1880, p. 359 (Tucuman).

Description.—Above and below nearly uniform chestnut, tinged with olivaceous on the crown and throat; narrow shaft-spots on the breast-feathers whitish, forming faint lines; beak pale horn-colour; feet bluish grey: whole length 11·10 inches, wing 5·5, tail 4·0. *Female* similar.

Hab. North Argentina, Paraguay, and Bolivia.

This large and powerful Wood-hewer is confined to the hottest portion of the Argentine Republic, and also inhabits Paraguay and Bolivia. White met with it at Oran, in the province of Salta, and writes concerning its habits:—"Common here in the dense forest, where their continued hard pecking at the lofty tree-trunks is very accentuated. Two or three at a time maintain a continued race up a magnificent clear stem as far as the branches, when they fly to the bottom of the next and do likewise."

This species is nearly a foot in length, the beak being about two inches long, curved and very powerful. The tail is stiff, being used as a support in climbing, and the claws are strong and sharply hooked. The colour of the whole plumage is nearly uniformly bright rufous.

---

221. PICOLAPTES ANGUSTIROSTRIS (Vieill.).

(NARROW-BILLED WOOD-HEWER.)

Picolaptes angustirostris, *Scl. et Salv. Nomencl.* p. 69; *White, P. Z. S.* 1883, p. 433 (Cordova). **Lepidocolaptes atripes,** *Barrows, Auk,* 1884, p. 21 (Entrerios).

Description.—Above, head and neck blackish, with oblong whitish shaft-spots on the crown and neck; broad superciliaries white, extending nearly to the back and broken at their lower ends into shaft-spots; rest of upper surface dull brown, brighter on the rump; wing-feathers pale obscure chestnut; outer webs and broad tips of primaries blackish; tail chestnut; beneath white, clearer on the throat; sides of breast and belly thickly marked with faint blackish stripes; under wing-coverts cinnamon; bill pale horn-colour; feet grey: whole length 8·2 inches, wing 3·8, tail 2·5. *Female* similar.

Hab. Paraguay and Argentina.

This is the only member of the genus *Picolaptes* as yet met with within the limits of the Argentine Republic. Azara found it abundant in Paraguay, and on this account named it the Common Climber, "*Trepador comun.*" In Buenos Ayres it is a summer visitant, appearing at the end of September. It is a solitary bird, never seen away from the woods, and invariably utters a loud melancholy cry when passing from one tree to another. It always alights on the trunk close to the ground, clinging to the bark in a vertical position, supported by the tail, and with head thrown far back in order to give free play to the extremely long beak. Having thus alighted, it progresses upwards by short hops, exploring the crevices in the wood for small insects, until it reaches the branches, when it flies off to the next tree.

Fam. XVIII. FORMICARIIDÆ, OR ANT-BIRDS.

The Formicariidæ, or Ant-birds, are another very important constituent of the Neotropical Avifauna, but are less generally diffused than the Dendrocolaptidæ, the greater number of the forms being restricted to the hot forests of South and Central America. In the Argentine Republic we meet with only four species of the widely-spread genus *Thamnophilus,* and of them only one appers to extend as far south as Buenos Ayres.

---

222. THAMNOPHILUS LEACHI (Such).

(LEACH'S BUSH-BIRD.)

Thamnophilus leachi, *Scl. et Salv. Nomencl.* p. 69; *White, P. Z. S.* 1882, p. 614 (Misiones).

Description.—Black, thickly and conspicuously spotted above with white or pale yellowish; on the wings the spots occur on the outer webs of the feathers; upper tail-coverts faintly edged with grey; tail black, unspotted; feathers on the lower breast and belly slightly edged with grey; bill and feet black: whole length 10·5 inches, wing 3·4, tail 5·0. *Female*: spots larger and yellow, on the crown lengthening to stripes and tinged with rufous; inner webs of the wing-feathers also spotted on their margins; whole under surface thickly spotted with pale yellow.

Hab. Brazil and N. Argentina.

White met with a single individual of this fine Bush-bird at Concepcion, Misiones. It was observed on the ground feeding on a swarm of black ants.

---

223. THAMNOPHILUS MAJOR, Vieill.

(LARGER BUSH-BIRD.)

Thamnophilus major, *Scl. et Salv. Nomencl.* p. 69; *White, P. Z. S.* 1882, p. 614 (Salta); *Salvin, Ibis,* 1880, p. 359 (Salta). **Thamnophilus stagurus,** *Burm. La-Plata Reise,* ii. p. 471 (Paraná, Tucuman).

Description.—Above black, upper wing-coverts crossed by a row of white spots; the outer rectrix on each side tipped and barred with white, the other rectrices, excepting the middle pair, spotted with white on their inner webs; beneath pure white; bill black, feet grey: whole length 8·0 inches, wing 3·6, tail 2·8. *Female* above rufous, beneath white, lower flanks and under tail-coverts tinged with ochraceous; tail wholly rufous.

Hab. Paraguay and N. Argentina.

White found this species common in the gardens round the town of Oran. He noted the iris as "crimson."

Azara's account of this species is prefaced by the following remarks on the birds of this genus known to him in Paraguay:—"These birds inhabit only the dense and tangled thickets, and never show themselves outside of their hiding-places, except for a few moments in the early morning and in the evening; but at no time do they perch high on the trees, but keep always within a few feet of the earth. They live in pairs; feed solely on insects caught in the bushes which they frequent, or on the ground, on which, however, the bird alights only to pick up its prey, and then returns to the twig to devour it. They are stationary, and fly only from one thicket to another. Many of the species have a similar voice or song, which is singular, powerful, and heard only in the love-season. The call is a trill of a single note rapidly reiterated, and loud enough to be heard half a mile away; the cry being accompanied by vibratory motions of the wings."

---

224. THAMNOPHILUS CÆRULESCENS, Vieill.

(SLATY-BLUE BUSH-BIRD.)

Thamnophilus cærulescens, *Burm. La-Plata Reise,* ii. p. 472 (Paraná, Mendoza, Tucuman); *Scl. et Salv. Nomencl.* p. 70; *White, P. Z. S.* 1882, p. 614 (Corrientes, Misiones); *Barrows, Auk,* 1884, p. 21 (Entrerios); *Salvin, Ibis,* 1880, p. 361 (Salta).

Description.—Above grey, black on the crown, and a patch of white and black feathers on the middle of the back; wings black, the coverts thickly spotted and the inner webs of the remiges edged with white; tail black, all the feathers, except the middle pair, tipped with white, the outer rectrix on each side has also a long white spot in the middle part of the outer web; beneath cinereous, white on the belly; bill and feet black: whole length 5·6 inches, wing 2·7, tail 2·3 *Female*: above earthy olive-green, yellowish on the crown; wing-coverts and tail as in the male; wing-feathers olive-brown; beneath yellowish olive-grey, brightening into ochraceous on the belly and flanks.

Hab. Paraguay and Argentina.

Mr. Barrows says concerning this species that it is "frequently seen, especially during the winter" (on the Lower Uruguay). A nest, taken November 24th, "was almost precisely like that of our Red-eyed Vireo (*V. olivaceus*), being pensile in the fork of a horizontal spray, only four feet from the ground. It contained three fresh eggs, white, with spots and dashes of light brown."

White says that these birds frequent the banks of streams, and have a deep bass call-note, rather mournful and slow.

---

225. THAMNOPHILUS RUFICAPILLUS, Vieill.

(RED-CAPPED BUSH-BIRD.)

Thamnophilus argentinus, *Burm. La-Plata Reise,* ii. p. 472 (Paraná, Tucuman); *Scl. et Salv. Nomencl.* p. 70; *Durnford, Ibis,* 1877, p. 183 (Buenos Ayres); *Barrows, Auk,* 1884, p. 21 (Entrerios). **Thamnophilus ruficapillus,** *Pelz. Orn. Bras.* p. 79; *Berl. et Jher. Zeitschr. ges. O.* 1885, p. 49.

Description.—Above olivaceous brown, tinged with rufous, brighter on the crown; lores yellowish white, superciliaries and sides of head whitish grey; wing-feathers dark olive-brown, the inner webs edged with ochraceous; tail black, all the rectrices, except the middle pair, slightly tipped and broadly spotted on the inner webs with white; beneath greyish white, the feathers from the lower throat to the belly transversely barred with black; bill and feet black: whole length 6·2 inches, wing 2·7, tail 2·6. *Female*: above the same as the male, except the tail, which is rufous brown; beneath whitish, tinged with ochraceous, and with scarcely perceptible greyish mottlings.

Hab. Paraguay and Argentina.

This species is very common in the eastern provinces of Argentina, and extends south to Buenos Ayres. It is a shy, solitary bird, found in woods and thickets along the shores of La Plata; and utters occasionally a singular low rasping note, its only language. The nest is a slight shallow structure placed in a low tree; the eggs are white, thinly spotted with reddish brown. Probably this species is to some extent migratory, as I have only observed it in the summer season.

Fam. XIX. PTEROPTOCHIDÆ, OR TAPACOLAS.

The Pteroptochidæ form a small but very peculiar family of Tracheophonine Passeres, mostly restricted to Chili and the south-western extremity of the South-American Continent, but also represented in the Andes of Ecuador and Colombia and in the high plateau of Central Brazil. They are ground-birds, remarkable for their large and robust feet with long claws, their strangely formed bills, and the elevated position in which the tail is carried in the living bird.

In the Argentine Republic four species of Pteroptochidæ are known to occur, only one of which, however, is a well-known bird. Two of the remaining three are recent discoveries, and the fourth a Chilian species, which extends over the Andes into the western borders of Argentina.

226. SCYTALOPUS SUPERCILIARIS, Cab.
(WHITE-EYEBROWED SCYTALOPUS.)

Scytalopus superciliaris, *Cab. Journ. f. Orn.* 1883, p. 105, t. ii. fig. 3 (Tucuman).

Description.—Nearest to *S. indigoticus* of Brazil, but without the white colour on the breast and belly, only the throat being clear white; superciliaries striped white; front and sides of the head and neck, breast and belly grey; rest of the upper surface, together with the flanks and crissum, light brown, with fine blackish cross-markings; bill blackish, feet light-coloured.

Hab. Sierra of Tucuman.

This species is one of Schulz's recent discoveries in the Sierra of Tucuman, where he found it west of Sauciyaca, frequenting the deep ravines.

227. RHINOCRYPTA LANCEOLATA, Geoffr. et d'Orb.

(GALLITO.)

Rhinocrypta lanceolata, *Burm. La-Plata Reise,* ii. p. 471 (Mendoza, San Juan, Catamarca); *Scl. et Salv. Nomencl.* p. 76; *Hudson, P. Z. S.* 1872, p. 543 (Rio Negro); *White, P. Z. S.* 1882, p. 614 (Catamarca).

Description.—Above, head and upper neck reddish brown with a fine white shaft-stripe on each feather, the stripes being most conspicuous on the crest-feathers; lower neck, back, rump, and wings greyish olive; tail blackish; beneath, throat and upper part of breast, grey, becoming pure white on the middle of the belly; sides of belly and flanks bright chestnut; lower part of belly and flanks and under tail-coverts like the back; bill horn-colour, feet black: whole length 8·2 inches, wing 3·3, tail 3·0. *Female* similar.

Hab. Western and Southern Argentina and Patagonia.

The vernacular name *Gallito,* or "Little Cock," by which this species is familiarly known in Patagonia, cannot fail to strike every one who sees the bird as very appropriate, for it struts and runs on the ground with tail erect, looking wonderfully like a small domestic fowl. In the neighbourhood of Carmen, on the Rio Negro, it is very abundant, and when I went there its loud deep chirrup, heard from every side in the thicket, quickly arrested my attention, just as the perpetual chirping of the Sparrows did when I first landed in England. In the interior of the country it is not nearly so abundant, so that man's presence has probably in some way affected it favourably. Its habits amuse and baffle a person anxious to make its acquaintance; for it scarcely possesses the faculty of flight, and cannot be driven up, but it is so easily alarmed, so swift of foot, and so fond of concealment, that it is most difficult to catch a sight of it. At the same time it is extremely inquisitive, and no sooner does it spy an intruder in the bush than the warning note is sounded, whereupon every bird within hearing hops up into a thick thorn-bush, where it utters every three or four seconds a loud hollow chirrup, and at intervals a violent scolding cry, several times repeated. When approached they all scuttle away, masked by the bushes, with amazing swiftness, to take refuge at a distance, where the loud protest is again resumed; but when the pursuer gives up the pursuit in disgust and turns away, they immediately follow him, so that he is perpetually encircled with the same ring of angry sound, moving with him, coming no nearer and never allowing its cause to be seen.

On three or four occasions I have seen one rise from the ground and fly several yards with a feeble fluttering flight; but when closely pursued in an open place they seem incapable of rising. They generally fly down from the top of a bush, but always ascend it by hopping from twig to twig.

The nest is made in the centre of a thorny bush two or three feet from the ground; and is round and domed, with a small aperture at the side, and built entirely of fine dry grass. The eggs are four in number and of a pure white.

The Little Cock is nine inches long, has long stout legs, and short curved beak. The colour of the head and neck is deep reddish brown, lined with white, the feathers of the crown forming a crest. The upper surface and wings olive-brown; tail dark; throat and breast grey; middle of the belly white, with a broad ruddy rufous patch on each side reaching to the belly.

---

228. RHINOCRYPTA FUSCA, Scl. et Salv.
(BROWN GALLITO.)

Rhinocrypta fusca, *Scl. et Salv. Nomencl.* pp. 76, 161.

Description.—Above uniform pale cinnamomeous brown; superciliaries whitish; tail blackish, except the middle pair of rectrices, which are the same as the back; beneath white, faintly tinged with cinnamomeous; bill horn-colour, feet black: whole length 6·5 inches, wing 2·5, tail 2·4.

Hab. Mendoza.

The original examples of this species were obtained by the German collector, Weisshaupt, near Mendoza. It appears to have been mistaken for the young of the preceding species, but is decidedly different.

---

229. PTEROPTOCHUS ALBICOLLIS, Kittl.
(WHITE-NECKED TAPACOLA.)

Pteroptochus albicollis, *Burm. La-Plata Reise,* ii. p. 471 (Mendoza); *Scl. et Salv. Nomencl.* p. 76.

Description.—Above chestnut-brown, obscure or slightly olivaceous on the neck and upper part of back; inner webs of wing-feathers blackish; lores, superciliaries, and under surface white, the breast sparsely, and the belly and flanks thickly marked with transverse spots of blackish brown; lower flanks and under tail-coverts chestnut-brown, obscurely spotted with black; bill and feet black: whole length 8·5 inches, wing 3·1, tail 2·9. *Female* similar.

Hab. Chili and Western Argentina.

Dr. Burmeister tells us that he met with this Chilian species in the water-courses near Mendoza, where it was seen running about on the stones like a Dipper.

Darwin has given us some interesting notes on this bird:—"This species is called by the Chilenos 'Tapacola,' or 'cover your posteriors.' The name is well applied, as the Tapacola generally carries its short tail more than erect, that is, inclined backward and toward the head. It is extremely common in Central Chile. The Tapacola frequents hedge-rows and the bushes which are scattered at a considerable elevation over the sterile hills, where scarcely another bird can exist; hence it plays a conspicuous part in the ornithology of Chile. The Tapacola is very crafty; when frightened by any person, it will remain motionless at the bottom of a bush, and will then, after a little while, try with much address to crawl away on the opposite side. It is also an active bird, and continually making a noise; these noises are various and strangely odd; one is like the cooing of doves, another like the bubbling of water, and many defy all similes. The country-people say it changes its cry five times in the year, which is according, I suppose, to some change of season. I was told that the Tapacola builds its nest at the bottom of a deep burrow." (Darwin, in Zool. Voy. 'Beagle,' iii. p. 72.)

Printed in Dunstable, United Kingdom

91230736R00141